Car Cultures

Edited by

DANIEL MILLER

Oxford • New York

First published in 2001 by
Berg
Editorial offices:
150 Cowley Road, Oxford, OX4 1JJ, UK
838 Broadway, Third Floor, New York, NY 10003-4812, USA

Berg is the imprint of Oxford International Publishers Ltd.

Library of Congress Cataloging-in-Publication Data

A catalogue record for this book is available from the Library of
Congress.

British Library Cataloguing-in-Publication Data

A catalogue record for this book is available from the British Library.

ISBN 1 85973 412 X (Cloth)
      1 85973 407 3 (Paper)

Typeset by JS Typesetting, Wellingborough, Northamptonshire.
Printed in the United Kingdom by Biddles Ltd, Guildford and
King's Lynn.

To Kathryn Earle – whose idea this was

Contents

Acknowledgments

The original idea for this book came from Kathryn Earle of Berg Publishers. Indeed I feel obliged to admit that when it was first put to me I was quite reluctant to follow it up. Although I had previously conducted some fieldwork on the use of the car in Trinidad, I am neither particularly fond of nor interested in cars and could think of almost no relevant literature. It was, however, the last point that persuaded me of the importance of such a volume, since in the case of every other aspect of material culture with an equivalent presence in the world the literature seemed to be at least to some degree commensurate with the importance of the object in question. I am inundated with books to read on the topic of houses, clothing and food for example. I realized that many anthropological colleagues had found the car to be significant within their fieldwork but it was as if it had not yet been made legitimate as a topic to focus upon. Even more surprising was the neglect of the topic of car consumption in other disciplines. On completing this book I now feel that it is a particularly significant contribution as it became clear as each chapter was submitted how much else might follow in terms of the comparative understanding of society through this particular vicarious route into social analysis.

I am very grateful to the contributors who have accepted the challenge with chapters of such high quality and insightfulness that I hope will be enhanced by their juxtaposition within a single volume. For my introduction I am grateful for comments and suggestions to Elizabeth Shove, Don Slater and Nigel Thrift. If the book is dedicated to Kathryn Earle it is also a reflection of the amount of work we have undertaken in partnership in publishing many other volumes both my own and by others. My experience of Berg Publishers followed that of several other publishers where I felt rather dehumanised as author, both powerless with regard to decisions made and facing an impersonal commercial agent. Berg, in my experience, while entirely forthright about their commercial requirements, as far as possible do not allow this to detract from a personal and involved negotiation with authors who are recognized to be usually passionately interested in the results of their labours.

List of Illustrations

Notes on Contributors

Michael Bull teaches media studies at Sussex University and has written widely on the role of sound in urban culture. He is author of *Sounding Out the City: Personal Stereos and the Management of Everyday Life* (Berg 2000).

Pauline Garvey teaches anthropology at the National University of Ireland at Maynooth. Forthcoming papers include 'Aesthetics and Practicality in a Norwegian Household' in C. Tilley, ed. *Material Culture and Social Identity* and 'Moving Furniture as Self-Introspection' in D. Miller, ed. *Material Culture and the Home.*

Paul Gilroy is Professor of Sociology and African-American Studies at the University of Yale. His publications include *Between Camps* (Allen Lane 2000), *Small Acts* (Serpents Tail 1994), *The Black Atlantic* (Verso 1993), and *Their Ain't No Black in the Union Jack* (Routledge 1992).

Birgit Meyer is a senior lecturer at the Research Centre for Religion and Society (University of Amsterdam). Recent publications include 'Translating the Devil. Religion and Modernity Among the Ewe in Ghana' (Edinburgh University Press, 1999) and *'Globalization and Identity. Dialectics of Flow and Closure'* (edited together with Peter Geschiere, Blackwell 1999).

Mike Michael teaches Sociology at Goldsmiths College, University of London. He is author of *Constructing Identities* (Sage, 1996) and *Reconnecting Culture, Technology and Nature: From Society to Heterogeneity* (Routledge, 2000), as well as numerous articles on the sociology of science and technology.

Daniel Miller teaches material culture studies in the Department of Anthropology at University College London. Recent publications include *The Dialectics of Shopping* (Chicago 2001), *The Internet: An Ethnographic Approach* (with D. Slater, Berg 2000), *A Theory of Shopping* (Polity 1998) and *Capitalism: An Ethnographic Approach* (Berg 1997).

Simon Maxwell is a researcher in the Environment and Society Research Unit (Department of Geography, University College London) and was previously a Senior Researcher in the Environment and Transport Studies Department at the London Research Centre. His interests are in transport and environmental governance, and qualitative research methodologies.

Tom O'Dell is an Assistant Professor at the University of Lund where he is employed in the Departments of European Ethnology, and Service Management. He is author of the book *Culture Unbound: Americanization and Everyday Life in Sweden* (Nordic Academic Press, Lund 1997). He is also editor of *Nonstop! Turist i Upplevelseindustrialismen*, (Historiska Media, Lund 1999).

Gertrude Stotz is senior anthropologist, at Pitjanjatjara Council Inc, Alice Springs, Australia. Her PhD was entitled *Kurdungurlu Got To Drive Toyota: Differential Colonizing Process among the Walpiri* (Deakin University).

Jojada Verrips is Professor of European Anthropology at the University of Amsterdam. He has written and edited a number of books in Dutch and is currently working on a book on the *Wild (in the) West*.

Diana Young is a senior lecturer at the Design School at Chelsea College of Art and Design and has trained as an architect and designer. She is also a Ph.D. student at the Dept. of Anthropology UCL. Forthcoming is *'Art on a String'* with Louise Hamby (Thames and Hudson).

Driven Societies

Daniel Miller

The Humanity of the Car

When I used to accompany my daughter to her school bus I often made up stories to amuse her as we walked. One was a description of the earth and its inhabitants as told by an alien examining us from a space ship above London. This alien had observed that the earth is inhabited by strange creatures called cars mainly with four wheels although some are great beasts with twelve wheels and some little creatures with only two. These creatures are served by a host of slaves who walk on legs and spend their whole lives serving them. The slaves constantly ensure that the cars are fed their liquid foods whenever they are thirsty and are cured if they have accidents: but the slaves also help in the reproduction and disposal of cars. The slaves are deposited in boxes set up almost everywhere a car wants to go and are always ready to be taken away as soon as the car makes up its mind to go somewhere else. Cars were never seen to go anywhere without at least one slave. The slaves build and maintain long and complex networks of clear space so that cars have little trouble travelling from place to place. Indeed the earth's creatures seems constantly pampered by their fawning army of slaves.

My point is that I can't think of any other object for which such a story could sound half so convincing. We may not be enthralled to cars, but the relationship of much of humanity to the world became increasingly mediated in the course of the last century by a single machine – the car. To such an extent that it is the car and its associated infrastructure rather more than the human that seems to dominate the landscape seen from the sky. Certainly there are many environmentalists who would be entirely happy with this perspectival conceit.

1

For them, but also in both colloquial and journalistic accounts, one does have a sense of the car as the sign of our alienation, in particular, from something which by opposition is seen as nature. The car is the villain that has separated us from the world and threatens to take over as we come to serve it more than it serves us.

Yet what this book will demonstrate is just how simplistic a concept such as 'alienation' appears to be when set against a relationship to cars which is not just contradictory but convoluted in the extreme. In this introduction I want to defend an approach which will remain critical but by starting from the opposite extreme. Perhaps for the first time this is a book that seeks to reveal and consider the evident humanity of the car. This should not be taken as some mark of adulation or defence of the car. To take seriously the humanity of the car must imply a perspective that examines the car as a vehicle for class, oppression, racism and violence, all evident products of our humanity. The car's humanity lies not just in what people are able to achieve through it, nor yet in its role as a tool of destruction, but in the degree to which it has become an integral part of the cultural environment within which we see ourselves as human. This includes both senses of the word – humanity as an expression signifying the totality of all people and humanity as a term that touches the specific and inalienable individuality of any particular person. The car today is associated with the aggregate of vast systems of transport and roadways that make the car's environment our environment, and yet at the same time there are the highly personal and intimate relationships which individuals have found through their possession and use of cars.

The title 'car cultures' is in turn intended to evoke the diverse, unexpected, sometimes tragic, contradictory humanity of cars; the taken-for-granted mundane that hides the extraordinary found in this material expression of cultural life. Indeed for all their freedom of imagination and resources the vast amount of television advertising which attempts to aestheticize the car within wondrous terrains of visual imagination tend to look bland and tedious against the fascination of what the observers who contribute to this book can document as the car's use in objectifying personal and social systems of value.

To speak of the humanity of cars is merely to foreground the proper difficulty of confining that concept to any simple definition. My foot could as easily be analysed as a technical construction of bone and flesh that provides me with means of mobility, as it could be seen as integral to my humanity. Quite often it rests on the pedal of a car, which in turn could be viewed as a mechanical achievement of metal and plastic that in turn gives me new means of mobility. But it is not so much of an extension beyond my foot, either physically or conceptually, for us to consider its humanity. Both foot and car are the basis of extensive entailments; both evoke the agency that mobilizes them, and the networks of relationships that walking and driving permit. Both have metaphorical and idiomatic reach. The foot is hopefully more permanent, the car more amenable to personalization and social appropriation (that is to say I suspect a whole lot more people think of their cars than of their feet as 'a kind of' person). But both fit that felicitous term 'second-nature' the habituated extension of ourselves that feels like nature in requiring no conscious mediation in their daily employment. Their humanity lies above all in the degree to which so many of us are socialized to take them for granted, so that we think our world through a sense of the self in which driving, roads, and traffic are simply integral to who we are and what we presume to do each day.

Nevertheless the mere idea of the car having humanity is not something that is easily accepted; indeed many readers may be revolted at the suggestion. For this reason Chapter Two by Diana Young, on the Pitjantjatjara of Australia, provides the ideal starting point for the wider intentions of this volume. It seems from the evidence she presents that it would be just as unnatural for these people to deny this manifest humanity to the car as it would seem unnatural for most people in Britain to acknowledge such a thing. In a very short period this Aboriginal society has assimilated the car into their material culture, so that today it is hard to hold a sacred ceremony without the car screening this from intrusive viewers, or to mark one's sacred sites without the car as the means to visit them. They follow and interpret the tracks made by cars in the landscape and they navigate cars through their sense of cardinal directions in ways that non-Aboriginal people are quite incapable of doing. In short the use of the car to facilitate one's phenomenological and cosmological relationship to the environment does not detract from that relationship which remains just as fundamental to Aboriginal life. The car has become more a means to resist alienation than a sign of alienation. Its materiality is no more of a problem than the materiality of the landscape itself which an extensive literature has documented to be foundational to Aboriginal societies.

This is why in Young's chapter, as also in Chapter Seven by Jojada Verrips and Birgit Meyer, a core problem is the car as a living and a dead being. The humanity of the car is expressed in the detailed issues raised by making the corpse of a car in some ways both analogous and symbolic of the human, and especially of a previous owner. The problem being that the dead carcass of the car, unlike other objects, is not so easily swallowed up by the soft sands of the Australian desert. In a similar fashion Verrips and Meyer reveal how many people's lives in Ghana are spent in the struggle to bring cars back to life, because a dead car is a threat to their ability to earn a 'living'. It is not surprising that there is a resonance with Christian concepts of miraculous resurrections such that a car mechanic needs spiritual as much as technical proficiency to keep a car on the road.

In Chapter Three, Mike Michael rests much of his argument on the problems that face us when we fail to acknowledge the integrity of such human–machine relationships. When our resolute dualism that separates off the humanity of people by contrasting it with the object nature of the inanimate prevents us from appreciating the hybridity of discourse and practice. The central point that emerges from his discussion of road rage is that the problems that arise from current discourses about such behaviour are exacerbated to the degree to which we oppose ourselves to the car and refuse to acknowledge its role in the formation and manifestation of such behaviour. Accounts of road rage are purified into two separate components – human behaviour and the technical effects of cars. By contrast a sense of hybridity allows us both to acknowledge and come to a more profound understanding of such genres as road rage. Road rage as discourse and practice can only be confronted from the depth of our now inseparable relationship with the machine in which we are socialized and through which we carry out our daily lives.

Those two chapters also constitute a helpful introduction to this volume in that they respectively represent two literatures within which it makes sense to treat the car from this perspective; a perspective which transcends an opposition of subjects and objects. Young is writing from the tradition of anthropological studies of material culture, in which the refusal of this dualism has been a foundational component of the discipline at least since the time of Mauss's (1966) influential work *The Gift*, where the basis of human sociality was seen as founded in the exchange of the humanity objectified in the objects given as gifts but also through his studies of technology and the cultural construction of the environment (Mauss 1979). This is a tradition that was fostered more recently by the work of Bourdieu and other theories of practice, and in the specific sub-discipline of material culture studies as exemplified in the *Journal of Material Culture*.

By contrast Michael is writing from within a strictly sociological trajectory which until recently had less interest in the specific issue of materiality, but which in the last few years has emerged into this particular light through the influence of science studies and most particularly the work of Latour. Latour's (e.g. 1993 and 1999) enormously influential critique of the dualism of science and nature has been extended through topics such as actor network theory, and work on monsters and hybrids to more generally challenge the distinction between persons and objects or between society and materiality. As a result although Young's paper is based on the classic anthropological techniques of ethnography and empathy with Aboriginal cosmology and social relations, while Michael's is an extended analysis of discourse and ideology that could in turn be described as classic sociological methodology, the final conclusions of both papers with respect to developing a new perspective on the car are very similar. Neither permits what may appear to be a common-sense separation of the car from a social or human context. Instead both offer an acknowledgement of a world of practice and discourse in which we can talk in terms of car cultures.

After establishing the humanity of the car as a starting point I will now address three approaches to studying car cultures. The first summarizes the dominant genres in which the car used to be presented to us in history and social science and considers the limitations of those genres. The second is a recent literature that is characterized as focusing on the problem of externalities, and the benefits and lacunae in that approach. Finally the chapters of this volume are considered as a series of studies of entailments, which complement and extend previous approaches to the study of cars.

Speeding Towards the Wrong Conclusions

In surveying the extant literature on the car within the social sciences and humanities (as against the more technical concerns of literatures such as transport studies which will not be addressed in this volume), I will adopt a critical tone. There certainly exist exemplary studies of the car (Brilliant 1989; Moorehouse 1991; Sachs 1984, to name but a few) but in a survey of the literature there soon emerges a sense of a dominant genre. This comprises two main tendencies. The first is a presentation of the history of car production and design which is understood largely in terms of the roles of major personalities and events. The second element is a presentation of the consequences of the car mainly in terms of aggregate statistics or universalizing traits.

These seem to have developed largely in the absence of any sub-discipline concerned with the car per se, so that the car appears most often as a case-study within the history of industrial production, design history and environmentalist critiques. By contrast, the mainstream disciplines that might have addressed what has been introduced here as the humanity of the car – that is, anthropology or sociology – have neglected the topic to a quite extraordinary degree, especially when compared to other examples of material culture such as food, clothing and the house. It would be very hard to exaggerate the disparity between the voluminous literature on those three topics and the lack of any comparable consideration of the car. There would be dozens of books equivalent to this volume already available for any one of those other objects.

The dominant genre tends to emphasize core events and persons. Most books start with the 'birth' of the car, the choice of which varies considerably from the steam-powered vehicle tried out in Britain in 1801 or the French combustion engine mounted on a coach in 1886 by Daimler (Wolf 1996: 67). But the key figure in the history of car production is always Henry Ford. We are informed of various crucial dates between 1 October 1908 when the first Model T went on the market for 825 dollars (Nadis and MacKenzie 1993: 3) and 1927 when the last of 15 million Model Ts rolled off the assembly line at a mere 290 dollars (Fink 1975: 67). Ford is seen as responsible for the US dominance in car production and ownership such that by 1930 there was 1 car for every 1.3 households (Nadis and MacKenzie 1993), although as O'Connell (1998) points out another factor was that the US had longer distances and lesser coverage by railways than for example the UK.

Given the generally anti-car tone of most recent writings, apart from being told how undemocratic Ford was (and the irony of the evidence that he was better known in the Soviet Union than Stalin in 1927 (Fink 1975: 71)), the other figure that is regarded as of equivalent influence in Europe is Hitler, who is credited with developing the modern autobahn system in Germany and promoting a Volkeswagen in emulation of the Model T in the US, though obviously at a later date. For the middle period of the century attention tends to turn to the history of car design and major figures such as Sloane or Harley Earle who influenced the development of companies such as Ford and General Motors (e.g. Gartman 1994) and shaped what became the dominant genre of car styling. Relatively little consideration is given to more recent events, such as the rise of the Japanese car industry, outside of the more technical and business literature where this is the focus of many studies.

Most of the recent texts move from these events in the history of car production to a flood of anti-car statistics for the present. For example, we learn that in the UK 'the initial production of every car involves 25 tonnes of waste (Graves Brown 97 25), that 11 million cars are retired annually in the US, and 240 million tyres junked each year (Nadis and MacKenzie 1993), creating 2 million tonnes of toxic waste. That in 1990, 420,000 people were killed and 9 million were injured globally as a result of car use (Whitelegg 1997) and that between 1960 and 1994 around 5 million people lost their lives in road accidents (for other examples see Davis 1992/3, Holtz Kay 1997, Whitelock 1971, Wolf 1996). These literatures on the history of car as a symbol of production and the current evidence for the car as a symbol of destruction along with a technical literature on areas such as transport systems, dwarf other writings about the car.

Occasionally the rather relentless tone of these trajectories is relieved by asides. These take two forms. One demonstrates the degree to which car-associated problems may have arisen prior to the car. We may think of our road system as a system for cars, but much of it existed prior to the existence of the car. For example there are several authors who note the huge increase in the extent and number of roadways in US cities at the turn of the century and the effects of horses on congestion. McShane (1994: 49) cites a gruesome practice of lighting fires under horses stomachs to get them to pull heavy loads, and Flink (1975: 34) records that 'In New York City alone at the turn of the century, horses deposited an estimated 2.5 million pounds of manure and 60,000 gallons of urine on the streets every day. Traffic was often clogged by the carcasses of overworked dray horses who dropped in their tracks during summer heat waves or were destroyed after stumbling on slippery pavements and breaking their legs. On the average, New York City removed 15,000 dead horses from its streets each year'. The other asides take the form of alternative histories of the car, for example (Schiffer 1994: 1) noting that 28 per cent of the 4,192 American automobiles produced in 1900 were electric.

More satisfactory and more heterogenous is the range of social histories of the car, although these too tend to be limited as a genre. Many start with the general condemnation of the car as initially an elite vehicle that exacerbated social distinction, a time when the archetypal driver seems to have been Toad of Toad Hall. In popular culture 'they were commonly revered as symbols of the good life aspired to by all, and turned up in Cole Porter songs, Fitzgerald novels and Hollywood movies'. (Gartman 1994: 60). Up to 1908 most car stories were about elites and appeared in the society pages (McShane 1994: 128). In some of this work one starts quoting literature that represents an elegy to a pre-car age, as in E.M. Forster's *Howards End,* but then moves swiftly to the contemporary sense of the car as symbol of destruction, quoting the sex and violence of J.G. Ballard's *Crash* (e.g. Graves-Brown 1997). We seem to suffer from what Sachs (1984: 173) calls the ageing of Desire which turns into the modern movement of disenchantment with the car. Earlier works such as Perkin (1976), and more recently O'Connell (1998), provide more sober histories that relate to wider issues such as class and changing patterns of work and leisure. In general this seems less true of fiction and entirely untrue of the cinema, where cars seem to have enjoyed a far more taken-for-granted centrality to the depiction of modern life, and genres such car chases, anthropomorphized cars, and cars as central to teenage dating movies are all standard fare. Popular culture does not seem to have undergone the same kind of reversal of attitude to the car that one finds in academic perspectives.

Even within social history there tends to be an emphasis on the consequences of the car rather than an empathetic account of car consumption in particular cultural contexts. There are exceptions such as Sachs's (1984) insightful history of the car in Germany. There are also particular areas of car use that are well served, such as an association with holidays where the popularity of the car receives much more empathetic and in-depth treatment for topics such as picnics and vacations (e.g. Lofgren 1999; O'Connell 1998; Sachs 1984: 150–160, and Urry 2000: 60–2). This is, however, rare since otherwise empathetic accounts of driving tend to be lacking from the historical texts.

The particular exception that perhaps best shows the narrowness of much of the rest of the literature is Brilliant (1989). She shrewdly picks on a key moment in the development of car cultures which is Southern California in the 1920s. She suggests the breadth that such histories ought to have, ranging from the car's relationships to popular cultures such as the movies, the context of its development in the relationship between class and new forms of mobility, the stance taken to the car by groups ranging from the law to the churches. As a result she can draw important conclusions. While at one level this is one of the most dynamic and formative periods and places where practices such as hitchhiking and joyriding are being given their 'modern' form, what emerges is how quickly these become established genres with normative behaviours and discourses associated with them. She extrapolates to contemporary Los Angeles where, despite what appears like car madness to the critics, for most people what can be observed is simply the struggle to make the intensive use and reliance upon the car livable with and mundane.

The most empathetic literature is perhaps not surprisingly that devoted to the car enthusiast, including Moorehouse's excellent study (1991) of hot-rod enthusiasts or O'Dell (1997) working with Swedish car aficionados; but there is an obvious problem in extrapolating from these studies to the more general and mundane contemporary relationship with cars which remains woefully unexplored. Overall then this is a literature in which the sense of the experience of the driver and the way the car fits into daily family life is hard to extricate from the statistics about car destruction or the key figures of car production. Gartman (1994), for example, provides a scholarly study of the development of car design, but while often invoking the agency of buyers always reduces this to assumptions about the possible impact of class conflict. Yet a history, especially a social history of the car that does not include the driver, is a fetishized history that makes the critique of the car appear abstract and distant from the humanity in which it is involved, even when the statistics are about death on the roads and pollution. The object and the subject are set radically apart, and in much of the recent literature they appear mainly as antagonists where humanity is always the victim. This has the problematic effect of taking even the violence of the driver outside of our sense of humanity, a point illustrated in this volume by Michael in Chapter Three. This is also why Sachs (1984) comes across as a more effective critic of the car than most, since the attack comes at the end of sensitive contexualizing of the car's social history. As Simon Maxwell in Chapter Nine of this volume makes clear, most ordinary drivers, even though they are fully aware of the power of dominant anti-car discourses, still express a much more ambivalent and nuanced discourse of the car that allows them also to acknowledge the benefits that clearly accrue to them from car ownership. As a result they confront their knowledge of the dangers and destructive consequences of the car not as an abstract discourse but as the contradictions of their lives.

Although there exists a developing literature on the social history of car consumption it is still very small beer compared to the role of the car in considerations of production. It is not just that business history is dominated by terms such as Fordism which take their point of reference from the car, but current studies of business and management make the car the archetypal object of business itself. It is this that makes so astonishing the degree to which recent textbooks on consumption will have chapters on homes, magazines and fashion but make virtually no mention of the car. It is likely that this division corresponds to wider distinctions that are associated with gender, but this seems insufficient to account for the extremes of this difference in attention to the car.

As a result of the paucity of a scholarly literature on car consumption, the critical accounts that dominate the treatment of cars in recent years are replete with sweeping generalizations as to the implications of the car for a more general 'modernity'. It seems that the absence of empathetic observation leaves room for all sorts of conclusions to be drawn that might with better knowledge and more sustained examination appear problematic. Rather than trying to summarize this literature I will try to exemplify it by taking just two of the most obvious examples. The literature that takes the car as a symbol or as the token for modernity tends to assume that contemporary car use is associated with an increased sense of mobility and speed. For example we learn that 'the subjective experience of automobility is invariably described in terms of pleasure, excitement, mastery and similar positive feelings' (Freund and Martin 1993: 97). Similar assumptions are often found in their crudest form in the psychological literature where we are informed for instance that 'at around 100mph, things begin to change. This is because the brain cannot cope with some of the rapidly changing signals coming from the eyes' (Marsh and Collett 1986) – which must be something of a surprise to the well-established and skilled sport of motor-racing.

The main problem with speed may be the rushing of academics to unwarranted conclusions. For example it may well be that the average person in Western Europe travelled 5 miles a day in 1950 and 28 miles a day today (*Guardian* 2/2/2000 G2, p7). But does it follow that we have a greater experience of mobility? Wolf (1996) notes 'If we incorporate all of those changes into the estimates, we would have to conclude that there was actually very little increase in mobility. The workers go to and from work five times a week (previously six), while the students continue to go to and from school and university. The average householder goes shopping three or four times a week. At the weekend, the average citizen makes one or two trips to the countryside or to visit friends and relatives and during the week, may go out again, for instance to the cinema. This was how it was in 1929, in 1950, and it is not essentially different in 1995'. Air travel has enormously increased the distances we go on holiday but most people are still taking a single holiday a year. As Sachs (1984: 109) notes, at the same moment that the car increases mobility and renders distance less problematic it is likely to affect also our sense of what distance has to offer and to threaten the mystique that arises precisely from the problem of its overcoming.

There is a similar problem with assumptions about modernity and speed. Again the statistics are quite clear about how much faster we tend to go in cars and other transport systems. But travelling at 80mph in a contemporary sealed car with good suspension simply doesn't feel like going particularly fast, which is one of the reasons it becomes so dangerous. Travelling at 500mph in a plane on a long-haul flight feels slow to the point of leaden – one needs to walk to the toilet just to retain some element of mobility. Contrast this with the social histories of travelling in an open car at 30mph when going out for a spin in the 1930s – that very likely felt like authentic speeding. That was the time when modernist intellectuals celebrated the experience of speed per se and commerce manifested that relationship to speed in popular genres such as streamlining. Thrift (1996: 264–6) suggests that the key period when the subjective sense of time was altered by a heightened sense of the possibility of speed comes still earlier, in the nineteenth century. Is buying a high-performance car but never actually using that high-performance an experience of speed? Equally for commuters stuck in traffic jams the simple association of the car with speed may seem more like a sick joke , , . as if . . . If one wanted a serious study of the meaning of speed one would do better turning to Weiss's (1996) analysis of the Haya of North West Tanzania. Here at least both speed and slowness are understood as core factors in people's moral and cosmological understanding of a range of processes from cooking to kinship. With that background we come to understand why, for example, coins and notes are nicknamed after cars and trucks as both cars and money become associated with what are perceived as the dangers of speed and the sense of 'heat' and disease that have become associated with it.

There are many similar examples of rushed conclusions when it comes to the experience of cars and its relationship to modernity. Should we assume for example the car is a 'cause' of the growth of suburbia when the desire for this compromise between town and countryside is clear from Victorian times? One of the most ironic consequences of the critical literature is in its treatment of gender. One would have expected some opposition to the gendered construction of the car as masculine. Yet in its desire to demonize the car this literature constantly asserts a relationship between the car with masculinity and violence that verges on the essentialist. In this volume the main example of transgressive and dangerous driving comes in Pauline Garvey's Chapter Six, on Norwegian women, while Simon Maxwell in Chapter Nine shows that what is absent from the critical literature is any sense of the centrality of the car to the mundane tasks of women. There is little empathetic treatment of the increasing association of the car with being caring parents, whether it is using the motion of the car to finally get an infant to sleep, or the sense at one stage in parenting that all one is doing is chauffeuring the kids to various friends' and activities.

Finally, the current literature has almost no grasp of the global reach of the car today except in matters of production and destruction. There is no sense that the car might be a different cultural form or experience among different groups. I have limited my survey to works in English, and cannot say if this is more generally the case, but at least in the English literature there seems no precedent for this volume which starts by refusing to privilege any particular region or group. Indeed even within the metropolitan areas, apart from some works on gender (e.g. Scharff 1991) there is little sense of the relativism required to understand the diversity of populations represented within those regions; commentaries in this volume such as Paul Gilroy's Chapter Four show that this needs to be our starting point and not simply a curious aside to a universalizing story of modernity.

Externalities

So far three varieties within the dominant literature on the car have been considered: the conventions of car history as a story of production and destruction, car social history, and the car as a trope in generalizations about modernity. In recent years these have been joined by a powerful new literature that has had a huge impact in developing a more effective critique of the car. This most recent literature is dominated by an approach I want to characterize as a concern with externalities. Although the word comes from economics it actually characterizes a much wider perspective on the car that is not particularly concerned with economic matters or the market. Within economics the clearest representation of externalities as a means of analysing the car comes from Porter (1999). His starting point is precisely the way the car fails to conform to more conventional economic treatments, because almost all the major costs involved take the form of externalities: 'almost all our automobile problems arise from the car's generation of external costs, when we get into our cars, we are prepared to pay the private costs of driving. But we ignore the external costs which, when added to the private costs, make the social cost of driving extremely high' (p.3). Porter bravely struggles to re-incorporate the car as an object of economic enquiry through costing each of these externalities. For example he tries to calculate the cost of global warming or highway safety, traffic, land use and auto disposal, though thankfully gives up when confronted by topics such as illegitimate babies conceived in cars (p.187).

Such an approach to the car fits easily within the growth of modern markets and processes such as auditing which have tended to follow a market model. This has recently been theorized by Callon (1998) who argues that for markets more generally the problem is determining what should be brought within the frame of calculation and what should remain within the outside world of externalities. What evidently and properly frustrates so many writers is the evidence that car culture has developed in such a way that drivers simply don't have to face the 'actual' costs of driving. Instead there is an almost complete break between what the critics see as private pleasure and public vice. This serves to exacerbate an already existing split between a transport literature that focuses almost entirely on the consequences of the car for the provision of infrastructure, where the main concern is to compare the car with alternative systems of transport such as the railways, and a design-and-style-focused literature that is concerned with the car as an object rather than with the consequences of its use.[1]

Within the critical literature on the car the term 'externalities' need not remain within the confines of economic or market analysis, as it has become the basis for a much wider perspective. In other words we need not constrict the term to its 'proper' use in economics, but rather see it as the perspective that has come to dominate most recent writings on the car. This is very evident in Davis (1992/3) whose problem, as he views it, is that what comes to be regarded as the advice of apparently neutral academics such as sociologists or safety experts reinforces an ideology that sees safety as an issue for its potential victims, while he wants to return the onus to the car and its drivers. For example, he argues that seat belts – by securing the safety of drivers in a crash – may make cars more dangerous for bicyclists and pedestrians. The problem lies in the perception and balance of risk. What such writing seeks to do is to bring back externalities and make them accountable, socially and not just economically. Many of the critiques of the car comprise histories which chart how externalities became external (e.g. Holtz Kay 1997).

By contrast Johnston in his book *Driving America: your car, your government, your choice* (1997) brings 25 years' experience as a lobbyist for the car manufacturers to bear on his topic. He seeks to demonstrate that the market can handle and cure most of the problems that the critics raise. Indeed for him consequences such as a rise in car accidents inevitably follow from all forms of government regulation. So here the car becomes the protected frame and the task is to prevent external agents and most especially governments from entering into that frame. While the critics see the cause of externalities as the legacy of a powerful car lobby that has sought to prevent the car being burdened with its own consequences, he sees the critics as a powerful elite grinding down the wishes and pleasures of the mass population.

Surely the 'classic' analysis of externalities has already become the paper 'The well-travelled yoghurt pot: lessons for new freight transport policies and regional production' (Boge 1995) which meticulously disaggregated every component of a pot of strawberry yoghurt made in Germany, from the metal of its lid to the source of the glue and the wheatpowder and other ingredients. For each of these separate elements the paper calculates the transport that had been involved in creating the final yoghurt pot. The point was precisely that the whole arena of transport had become an externality to the purchase of mundane goods such as yoghurt, so this article very effectively reintroduces the 'true costs' of transport and its effects on the environment as an audit of yoghurt production. Much of the literature on externalities has now become an attempt to generate a larger audit of the car in terms not so much of financial costs but of environmental, social and ecological 'costs'.

A major advantage of this approach via externalities is that it raises consciousness as to the political context of car development, which in turn helps us see the importance of particular forms of political intervention. The history of the car has always depended upon an important articulation between a kind of micro-politics of civil society and the state. Sachs makes this very clear for Germany. He notes that in 1912 there arose a debate as to whether pedestrians – who, when confused by the new cars and unsure what to do, propel themselves in front of a vehicle – are to be blamed for their own death, or whether the responsibility is the driver's, or whether simply the conditions make such events inevitable, and in which case should the pedestrian be forced to concede the right of way to cars (Sachs 1984: 30–1). But civil society comes to be fully complemented by the state, as when Hitler confirms the cessation of space to the car with a vast programme of highway construction. For Hitler the car promised greater access and thus greater unity as a means to construct an ideal of 'One People, One Reich, One Fuhrer' (ibid. 53). But before conclusions are reached too quickly it is the car again that is central to a post-war development in Germany of social democracy and mass participation in the 1970s. What can be concluded is that the consciousness of the car as a problem of responsibilities and progress almost always involves a state whose laws and road-building programmes establish both infrastructure and many of the terms within which any car consciousness or discourse can operate.

So it is not surprising that political literature perhaps best explores the academic consequences of this situation, because to a striking degree these debates about the car cut to the heart of core questions about citizenship and contemporary civil society. As Rajan (1996) notes, the problem is that the car has become much more of a right than a responsibility: 'to drive and operate automobiles has become almost the inalienable right of every individual to achieve goals and purposes efficaciously across a space specifically engineered for this purpose' (6–7). What is required, he argues, is a 'civil society of automobility' (p.16). The political debate based on externalities recognizes what Urry (2000: 59, see also 190–3) complains that sociology has failed to recognize: 'The car's significance is that it reconfigures civil society involving distinct ways of dwelling, travelling and socialising in, and through an automobilised time-space.' One of the most instructive works is Dunn (1998), because at one level his is a straightforward critique of the mass of critiques of the car. Dunn pillories what he sees as the simplistic and unrealistic rhetoric of the attack, but his conclusions follow from this debate about externalities. He returns to a politics in which he insists it is possible to put the onus back on to Detroit to make cars safer and less polluting rather than depriving the drivers of their cars when they have no serious alternatives.

It is hard to exaggerate the importance of this literature on externalities. What is achieved is an opening-up from any simple focus on the car as merely a relationship with its individual owner, forcing us to raise our sights and imaginations to bring in all that the car implicates: aggregate effects, landscapes of roadways, patterns of work and patterns of leisure. In short a concern with externalities goes a long way in taking us from the car per se to the consideration of car culture. It is also crucial for future consideration of the car. At a time when companies are developing the 'intelligent' car, and governments the 'intelligent' road, the relationship of responsibility between commerce, the state and the private driver is only going to become more problematic and make the issue of civil society more acute. I therefore do not want to detract from its contribution. But nevertheless much of what is important in the present volume lies precisely in the manner in which most of the chapters here take a quite different and complementary route to the same issue of relating cars to car culture and thereby highlight much that is missing in those approaches that focus largely on this problem of externalities. There are three main advantages to this approach over that of externalities, which result from the literature on externalities tending first to reductionism, secondly to ignoring the need for a scholarly account of car consumption, and thirdly to a failure to problematize the basic question of what the car is.

The first limitation to the study of externalities comes from the degree to which it is influenced by the source of this model in economics. The positivism which is characteristic of the discipline of economics tends to result in the inclusion of externalities only to the degree to which they are measurable. The study of car safety, for example, tried to remain for some time under the umbrella of what was called Smeed's law (a Professor of Transport Studies at UCL) which claimed that the relationship between motorization and fatalities was irrespective of time and place (e.g. Whitelock 1971:xi, Davis 1992/3: 38). Cultural distinctions of all kinds tend to be downplayed. As a result, as Whitelock (1971) notes, the critique is effective in determining the impact of drinking on driving but rather less good on deciding what lies behind one of the principle causes of accidents, inappropriate overtaking on country roads. This is because such overtaking involves much more subtle questions about the nature of driving that cannot easily be reduced to a measurable variable such as the level of consumption of alcohol.

The extremes of this approach could be seen within work by Marxist economists such as Baran and Sweezy (1968: 139–41, quoted in Slater 1997: 136) who try to separate the car as a purely rational object of use, as against the high costs of making it part of commodity culture associated with advertising, style and other examples of what they see as unnecessary accoutrements. The effect of this approach – to strip down the car to measurable elements of function – is to make culture itself an externality, as though rationality is something founded only in function in its narrowest sense. Obviously in a book called *Car Cultures* the externalities that may matter are a whole swathe of social and cultural entailments that the current literature is not well equipped to either study or consider. What is explored in this book is an intimate relationship between cars and people that so far has paradoxically remained an externality within a literature that sees itself as dedicated to incorporating all other externalities.

This is particularly a problem when the literature shows that so much depends upon complex issues of moral and political discourse and a wider sense of risk. What Chapters Four, Five and Six in this volume (by Paul Gilroy, Tom O'Dell and Pauline Garvey respectively) strive for is something missing in the prior literature. All three chapters commit themselves to relate the car to its wider context in political economy – both commerce and the state – but in such a manner that this sheds light upon, rather than being opposed to, the more personal and involved relationship between values of particular groups of drivers or passengers and their cars. The problem for the study of car cultures, as of culture more generally, is to retain the link between the micro-history of ethnography of experience and an appreciation of the way these are shot through with the effects and constraints of acts of commerce and the state.

The third way in which the study of entailments differs from that of externalities is the different attitude to the basic question of what a car is. Ironically the one thing the study of externalities fails to do is to problematize the car itself. It almost inevitably assumes we know what a car is and that the problem is only to acknowledge all those consequences that have become disconnected from the car as their point of origin. In this book, by contrast, there are many examples in which we start either from some particular aspect of the car such as its audio system in Michael Bull's Chapter Eight or some larger association of car use such as gender in Gertrude Stolz's Chapter Ten and then gradually work our way back towards some new sense of what the car seems to be when viewed from that perspective. In this volume, then, the car is the conclusion of our work, not its premise. If the car is understood to be as much a product of its particular cultural context as a force then it follows that, prior to an analysis of that larger cultural environment, we cannot presume as to what a car might be.

What is a Car? Starting from the Upholstery

To make this last point clearer a case study is presented here in which the nature of the car itself emerges out of a process of academic enquiry that started with an investigation into car upholstery, and tries to end with insights into normative cultural values. In this case the study took place in Trinidad. (What follows is an abbreviated account of Miller 1994: 236–45.) Since in most respects I don't particularly like cars or pay them much regard, I never chose or intended to study them. I came to the topic of the anthropology of the car because I was completely overwhelmed by the presence of car upholstery. When I chose to undertake fieldwork in the town of Chaguanas in Central Trinidad in 1988, I didn't even know that car upholstery existed. I was somewhat perplexed to find that the part of the town in which I lived was completely dominated by this profession. Out of 176 commercial establishments in the area 38 were solely concerned with cars. Although this includes garages and car-part specialists these are dwarfed by three stores devoted to car upholstery, whose owners have also become wealthy enough to own much of the commercial property in the area. Apart from domestic upholstery at Christmas the trade is largely car upholstery, and these three merely dominate a large number of smaller car-upholstery firms in the area, making this probably the single leading commercial concern of the town.

In the recession period when I started fieldwork, upholstery was dominated by repairs to vehicles such as taxis. During the previous oil boom most new cars had had their first outing to these upholsterers, who might change everything from car seat to trunk (boot) with designs such as fake snake skin or a black leatherette streaked bluish and silver, marketed under the title of 'New York by night'. The dashboard could be upholstered, cushions added, and the whole complemented by a variety of paraphernalia such as perfumes, religious icons and stickers. Cars could also be feminized, for example, with heart-shaped satin cushions with projecting pink frills and central flower designs. The degree of such personalization was satirized by a journalist who recorded his removal of the accretions to his new second-hand car (*Trinidad Express* 20 Oct 1988): 'The tiger-skin covers came off on day one, as did the red plastic steering wheel cover; as did the little duckie. The white JPS emblems made it to day two, but no further; nor did the "I love my Mazda" sticker. Presently slated for retirement are: the dashboard heart that lights up in red with the words "love caressing"; the pair of little green bordello cabin lights; the red hyphen lights above the front number plate; the fog lights inscribed Denji; and at least one of the three antennae.'

The commercial centrality of car upholstery in turn forced me to acknowledge the objective evidence that, notwithstanding the relatively short time most people in Trinidad had had access to cars, cars were far more integral to identity and daily life than I had been familiar with in Britain. Some of this relationship was economic: for example, the importance of 'pulling bull', whereby drivers whose cars are not registered as taxis nevertheless use their cars for taxi work employing a hand gesture to tell potential passengers of their role. Approximately a third of all lifts I had in cars in Chaguanas during my fieldwork came from such private cars. This was in addition to the high degree of ownership of actual taxis. But much of this intimacy with cars could not easily be related to any economic effects. I soon learnt that individuals were located more often through the car parked in front of a house than by the house number. The local press constantly spread scandal and innuendo through reference to car ownership as in 'The leader has a nickname which resembles that of a popular large local fruit, and he drives a taxi which is neither too dark or too light' (*The Bomb* 21 December 1990), or it will talk of an AIDS victim 'whose husband drives a Mazda'. The retailers I studied, irrespective of what they were selling, routinely decided their expectations of a particular customer entering their shop on the basis of the customer's car, and so I would be told that a Laurel driver bought this but a Cressida driver would not buy that. The identification of persons by their associated cars is then not the exception but the norm of daily social discourse.

The situation was not as extreme as in Bermuda where Manning (1974) reports parties or even funeral notices being given on the basis of a person's car number plate because these were better known than their names, but it would also be true for Trinidad that gossip depends a great deal upon where a particular car has been seen and it is commonplace to infer things about the owner. Most people recognize a large number of such number plates. Car parts were also central to the sexual innuendo that is found in conversation and also calypsos, such as where a mistress is referred to as a 'spare tyre' and girls who pick their men by the bodies of their cars rather than that of their drivers as 'gasbrains'.

I also had to get used to an unwillingness to walk, as evidenced in the queue of cars in front of the school gates as each waits to put down or pick up at the exact entrance, and the tendency to have conversations from car to car often to the dismay of those parked behind in the traffic. Not surprisingly this was also reflected in car care. Anecdotes were common about neighbours who wash the car at least once a day, twice if it has rained, and who pay particular attention to the area within the treads of the tyres, and the sheer tension when a car door is slammed. Clearly then to understand this centrality of the car I would need to follow its further entailments in both social and material domains.

My next observation was that while the upholsterers are essentially concerned with car interiors, in the same high street are found other shops which are devoted entirely to car exteriors, such as the tinting of window glass and the adding of stripes to the exterior or wheel hubs – a key fashion item that at the time was shifting from metallic to white. I then surmised that the interior and exterior activities seemed focused upon two distinct groups of customers. While a car with fake tiger skin and pink plush interior might also have flashing lights on the exterior, in general retailers distinguished between a 'cool' look emanating from stripes and tinting and the 'flash' look embodied in some of the more outrageous upholstery. In turn, as so many dualisms in Trinidad, these are presented first in terms of ethnic stereotypes. A conversation described the 'Indian with gold on his fingers and hair greased back who wants crushed velvet upholstery but can only afford short pile acrylic but spends ages brushing it the right way' but also the black dude with his mistress projecting loud music from a car with tinted glass and stripes on the exterior.

Based on this initial separation of inward-looking and outward-directed car transformations I could link car upholstery with the extensive use of upholstery in the area of home furnishing. Such upholstery is generally evocative of interiorization as a value expressed in the living room, which I had anyway intended to study. The use of common elements such as the colour maroon and a similar employment of plastic-covered seats helped link the two sites. Their association was extended further when I interviewed the other upholstery industry in the area which are the funeral parlours, where coffins and the more expensive and luxurious caskets are almost invariably lined in deep buttoned upholstery. This led to the conclusion that the car was expressing a contradiction. On the one hand some treatments of upholstery turned the car to the values associated with the interior and the family (continued aesthetically even after death), while on the other hand an emphasis on display turned the car in the opposite direction towards a concern with individualism and mobility associated with the world outside the home.

As I followed these leads more widely I found the same dualism apparent in many other fields, which led ultimately to a representation of Trinidadian culture in terms of an opposition between two basic principles. On the one hand what was termed 'transience' tended to an orientation to the present and was focused on the public arena, where things are brought out into the open. These values were most fully expressed in the festival of Carnival. By contrast there coexisted another set of values termed 'transcendence' that was concerned with pasts and futures and also with interiors such as domestic life and the family. This set of values was most fully expressed in the festival of Christmas.

In turn I argued that many parameters of social life that tend to be seen in dualistic terms – including class, gender and ethnicity – may be derived from this foundational contrast between transience and transcendence. This had evident implications for the analysis of the car. To return to the problem of the dominance of car upholstery: the majority of people in this town are of South Asian origin and the male Indians are the dominant car-owning group. I argued that the car had become a key expression of their emulation of a sense of freedom which is associated with the lifestyle of African males, but constrained by modes of family life and attitudes to possessions associated with Indians. For example much of the fanaticism of car care seemed aggressively aimed at other members of the family who might feel they had a 'traditional' right as family members to use the car. The car is a substantial possession which usually 'delivers' on its promise of greater autonomy and freedom, but this does not necessarily mean that the driver abandons one set of values for another. The transference of domestic upholstery to the car seemed to be directed at retaining values asserted in the world of the domestic interior while enjoying experiences which in Trinidad are often seen as in direct opposition to these values.

While another analysis might have interpreted the car thereby as an outer expression of a core ethnic division, I saw the ethnic division as itself a manifestation of a foundational conflict in values that I had come to understand through the analysis of car upholstery. This conflict could also be found manifested in the way stereotypes of gender and class were being developed. This seems to me to indicate an advantage to the ethnography of material culture. Instead of assuming some social parameter such as class or gender and reducing one's observations to reflections of those, one works in the opposite direction. It is by finding out what a car is that one finds out what normative culture has become and how social stereotypes are thereby generated. I started with the unexpected observation that the town where I chose to live was dominated by the car-upholstery industry and end in understanding that this is because the cultural values of contemporary Trinidad are objectified precisely in car upholstery. The car itself then becomes understood not as a starting point but as an object whose presence can be comprehended as part of the movement from the study of car upholstery to generalizations about Trinidadian values. Through the study of entailments we come to a conclusion about what cars in particular contexts have come to be.

Entailments

The study of Trinidadian car upholstery illustrates how the study of entailments develops into the study of car culture. Indeed in this case my claims to any understanding of culture itself, as the play of normative values upon individual practice, arose through the problem of understanding the car. This in turn gave me my sense of what the car in Trinidad had become. But there are other chapters in this book that better exemplify the other ways in which the study of entailments complements the study of externalities. The first of these is the need to find a new way of connecting the larger aggregate effects of the car and the involvement of wider forces such as the state and the market with the more personalized and intimate relationship between cars and their users. This is something the approach to externalities conspicuously fails to do. Chapters Four, Five and Six in this volume help bring out the larger relationship to the state, to the political economy and to historical process that go well beyond the links made within my Trinidadian example.

For Gilroy in Chapter Four, the intense and intimate attachment between a particular group identity that US blacks developed with and through the car, and which he shows to be richly seamed in music and other aspects of black popular culture, can only be understood in terms of the larger history in which it was the market and wider social exclusion that made the car such an object of desire. A dialectic of overcoming that negates this initial separation is the fertile ground for this subsequent cultural efflorescence. More problematic is the extent to which the subsequent car culture acts or fails to act as a defetishism of the relationship between a history of oppressive labour in car production. This may resonate with a deeper and largely ignored ambivalence about the relationship to the car. What Gilroy excavates is the sense that, if one can see the repressed behind what seems at one level to be so overwhelmingly expressive a relationship between people and cars, then this could open out a more profound sense of the black route through US history. This history brings back into view, but through a critical lens, the centrality of consumer culture that is constantly reiterated in black popular culture. As such it exposes the limitations of more conventional history and narrower conceptions of the 'politicized' that fail to comprehend the implications of consumption for core alignments and misalignments as here between race and class. As has also transpired in other chapters, Gilroy highlights the centrality of consumption, and particularly car consumption, to the emergent consensus as to what areas of life we are prepared to acknowledge as politicized.

O'Dell's Chapter Five study follows well from that of Gilroy since in many respects his chapter documents the way these tensions within

US identity have become exported partly by virtue of the way the rest of the world has come to regard the car itself as a core symbol of American culture. In this case we find the car thereby becoming a prime element in the development of the Swedish sense of itself within the movement to modernity. On the one hand it could be the vanguard of a modernity that Sweden in general was keen to emulate and develop, but only if it could be an acceptable Swedish inflection. This contradiction is fully exploited in the 1950s by young greasers (*raggare*) who brought out precisely the sexuality, freedom and danger that their parents wanted to strip away from the American version of modernity. So in one sense the *raggare* seem to require what for them was the greater authenticity of the 'really' American car. But as O'Dell traces in detail, the way this relationship was played out within the larger compass of generational and class conflict in Sweden, it becomes clear that in their own way the *raggare* had become just as localized and specific to their own agendas and agency as the forces that tried to repress them. O'Dell's analysis is reminiscent of one of the founding papers in the modern study of consumption where Hebdige (1981) showed how motorbikes and motorscooters could occupy a similar position as both general signs of modernity but in their case objectifying rivalrous versions of the proper route to the appropriation of that modernity.

Garvey's Chapter Six emphasizes the still more paternalistic regime represented by the Norwegian state. This in turn clarifies the role of the state in determining the subsequent form taken by the relationship between a segment of the population and the car. The Norwegian state in a mix of Puritanism and paternalism tended to regard the development of the car as something to be feared and highly controlled, for which individual users would have to apply for a permit. In effect they thereby bracketed the car with other substances such as alcohol, which is still regarded with deep suspicion, and where an act such as drink-driving is regarded with particular horror. Seen from the opposite perspective of an ethnographic engagement with working-class women, Garvey observes the degree to which such women seem in turn to regard the car in much the same way as they regard alcohol. They are both ideal vehicles for transforming their behaviour in what locally is experienced as quite a radical 'flip' from a highly normative order to what would locally be regarded as highly transgressive disorder. As a result a particular genre of dangerous or 'deviant' driving becomes a clear equivalent to 'drunkenness' in the repertoire of their forms of practice.

In these three chapters, then, the problem of articulation between car culture and the wider political economy is constructed through a sensitivity to issues of power, particularly as expressed in forms of autonomy and cultural construction that as Gilroy indicates even where they may not effect any empowerment are certainly expressive of the sense of disempowerment. The car as working-class culture is either potentially re-thought back to the place of the working class in car production or, as for many of the women involved, it is swept into projects of sexual and other forms of freedom. If, however, these three chapters achieve something that was missing in the prevailing literature in terms of such an articulation, they in turn bring out the need to not just understand but find a way to express the intensity of the subsequent relationship to the car. It is this highly visceral relationship between bodies of people and bodies of cars that forces us to acknowledge the humanity of the car in the first place. And this is not something that emerges sufficiently either from the current literature or from a case such as the car upholstery in Trinidad.

It is the car enthusiasts who most clearly and fully express this relationship, as can be seen in the work of Moorehouse (1991) and O'Dell (1997). These are the people who seem to cultivate their cars to the degree that one feels the transformation of the car body is a vicarious expression of their sense of bodies more generally. This is brought out well in a recent journalistic account of cars in weekend car meets in contemporary Britain, and one car in particular that began life as a Ford: 'the wheels were snug to their pink rims, which flowed silkily into the wings, the doors, the side skirts, which skimmed the ground. The lines were fluid; everything about the car was smoothed and rounded, scooped in, curved out. The wing mirrors hugged and flexed to their doors. No sharp angles, no nips and tucks. No superfluous chroming or plastic.' Inside 'The poles were painted lime green, and beneath them, where the back seat should have been, were two of the biggest built-in car speakers I had ever seen . . . The bass started shaking my loose change. And this was over the sound of an engine, which was revving and booming like Concorde with a cough.' (Sawyer 1999: 31–2).

Behind this more extreme relationship Sawyer highlights the more mundane but perhaps more central role of the car at the stage of such teenagers' first aspirations, thereby linking the car enthusiasts of O'Dell's chapter to the more general relationship found among US blacks or Norwegian women in the other two chapters: 'a car is the only one of your dreams you're likely to see come true . . . advertisers understand that you may not ever run a company or own a mansion or have rails of couture clobber or go to Bali for your holidays – but, for a few hundred quid you, too, can buy a car' (ibid: 26). The car in general does not disappoint but is the means to realize that which it has promised, as she recalls: 'There's not much that can beat that feeling, even now, of music, and motor and people yapping nonsense and knowing that there's stuff to come, that the evening's not even started, that you and this fantastic machine are getting closer, being drawn in, towards the lights and the dark and the possibilities . . . Eventually driving like that leads to stopping. Stopping in some lay-by and leaving the tape on and getting into the back seat and pushing the front seats up against the window and grappling with zips and tongues and twisting clothes and limbs because there's nowhere, really, for you to go, because your parents are in and the clubs are all shut and there's only the car that can give you the space to discover what you are and who you like and who you are and what you like.' (ibid: 29)

What remains missing is the way this initial more celebratory relationship remains intimate and intense even when it also becomes that much more mundane and banal, as for middle-class and middle-aged men or housewives. For this reason perhaps the core to a new car literature ought to be in the experience of traffic. Yet this is perhaps the biggest of the lacunae. Traffic is central to the literature on externalities, but it is always as a simple trope of alienation, never an expression of the tightness of our relationship to cars. It almost comes as a relief to read Parker (1999: 31) who perhaps comes closest to the single most common experience of car use for most people today. As he notes, 'London drivers fear the morning peak, and the evening peak, and the school run rush hour, and the West End theatre rush hour. They fear the Saturday afternoon traffic and Sunday night traffic. And they fear the prospect of leaving a good London parking spot, when a car fills the space they have left, they feel troubled and adrift, regretting their recklessness . . . They fear the thousands of streets whose parked cars make it just too narrow for two cars to pass, and where they must play complex, draining games of oscillating generosity and aggression . . . They have a great fear that they are losing a race. The race is with an imaginary car that set out from the same place at the same time, but then did not get stuck behind the 31 bus, did not miss the lights, did not make that unforgivable lane error on Commercial Road. This car is way ahead.'

Two of the chapters in this book are particularly concerned both to convey and to analyse the intensity of ordinary relationships with the car that extend beyond the high-pitched charge of youth. Chapter Seven by Verrips and Meyer is particularly important in taking us away from a Euro-American history and also from an assumption that this intensity can only come through the act of car consumption. In their discussion they give us a sense of the humanity of the car as integral to the people who possess cars that is just as powerful and effective as in Chapters Two and Three by Young and Michael. What is portrayed is a kind of eternal struggle in which the individual can only gain his or her own life through his or her struggle to keep alive this machine which not surprisingly is conceptualized in relation to various spirits and forces that mediate between human and other forms of agency. But in their case it is not the experience of driving but that of repairing the car that is central to this relationship. It is not just that we are taken down from the world of Detroit's 'Fordist' systems to a tiny workshop in the backstreets of Ghana. It is also, as the authors conclude, that the very act of maintenance is here not separated out from the rest of life as the proper task of professionals or even workers: its integrity to daily life where almost everyone has some view or experience of car maintenance, produces something of that sense of auto-citizenship that the political discussion of externalities is trying to resurrect, though here in quite a different context than that literature could ever envisage.

Chapter Eight, the one most clearly directed towards the intimacy of our relationship with cars, is Michael Bull's discussion of the audio world of cars. This seems to follow from what Parker expresses as the ordinary sense of frustration and failure that drivers experience. Bull shows that this produces an extraordinary ability to cut oneself off from the world while at the same time, for safety reasons, one is aware of every millimetre that separates the car as one's extended self from others. In this account which (as that of Parker's) I find unnervingly close to my own experience, the car has become far and away the/my primary place for loud and utterly absorbing music. Bull's contribution is particularly important because it focuses upon that aspect of the car that is most absent from academic accounts though vital to popular culture, which is the experience of driving. Indeed by picking on the very particularity of the effect of the sound system he explores an avenue that would be productive for all aspects of car use. It is the paradox by which we can negotiate a few centimetres of rushing space with hardly a glance and yet people pick their noses at traffic jams as though the windows of the car were a barrier rather than a view onto their interior space. Bull's paper shows how sound has effectively

colonized this niche provided by a kind of ambivalence about consciousness, so that one can as it were both find oneself and lose oneself
in the intensity of music and voice in the car in a manner that makes
this more of a home space than the actual house. The sound as much
as the car creates a particular mode of space and experience, that can
be constructed even within that relentless and enforced sense of
alienation that Parker describes as traffic.

From the re-articulation of micro- and macro-perspectives found in
Chapters Four to Six and the sense of the intimate relationship between
people and machine found in Chapters Seven and Eight, we are in a
better position to see the way an approach to entailments complements
and extends the approach to externalities. But the intention is not to
attempt to replace one with the other but to conjoin the benefits of
each, and there are a number of ways in which this can be achieved.
One of the positive developments in the recent writings about cars
has been a preparedness to explore a wider range of entailments each
of which seems to spin off its own history and sometimes eccentric but
promising perspective on wider issues. These can make for fascinating
reading; for example, Jennings (1990) and Wachs and Crawford (1992)
include articles on subjects such as the history of garage architecture,
how trucking became central to US vernacular culture, the forms of
roadside advertising and the ideological significance of route 66 from
Chicago to Los Angeles which among other things influenced Stein-
beck's *Grapes of Wrath* as the route for the dustbowl exodus. Through
such studies one can glimpse the range of social and material entailments of the car that might be explored by a more adventurous and
imaginative academic pursuit, though they also suggest other areas such
as the experience of parking that have remained neglected.

The point that may also emerge from the range of such studies is
that we cannot presume how the stance to the car should be divided
up. In the literature on externalities we can see clearly divisions between
the treatment of the car as object, the private experience of driving
and the public concern with the consequences of the car. But from the
point of view of entailments the situation is more complex. What
counts as public and what as private is too bound up in larger histories
of ideology discussed in Chapters Four to Six and too unclear in particular ethnographic encounters as reported in other chapters within
this book. What the study of entailments suggests is that we need to
transcend oppositions between say transport studies and design studies,
and allow a focus upon the car to transport us intellectually between
otherwise separate academic destinations.

As things stand we might expect to come across an empathetic account of a driver, but it is very rare to find one based on the study of transport infrastructures. The same Parker (1999: 14–17) who writes so eloquently about traffic also makes an attempt to do just this with a rather inspired example of entailments in his story of a traffic signal engineer dealing with the notorious Hanger Lane gyratory system, where a major London orbital meets a major London radial used by around 8,000 vehicles an hour. The tale of how this engineer found a spare seven seconds in the sequencing of the traffic lights that could set the traffic moving again becomes positively heroic. Parker also manages a couple of pages on the history of painting road lines (ibid: 19–20) and speculates on the potentials of city gridlock.

To fully re-engage the literature about externalities with that on entailments however, we need to return our studies to the debates about ethics and futures which generated so much of the current literature on the car and keeps it so committed. In his Chapter Nine, Maxwell attempts to do this by returning these debates to the drivers. Since these debates are so vocal and disseminated so widely in the media, they become in turn the backdrop to the reflexivity of the drivers themselves, which is precisely what the critics of the car intend. But what Maxwell reveals is a further consequence of this reflexivity in a context where – as has been argued throughout this introduction – the critical literature has cut itself so far off from the experience and concerns of drivers that it becomes as much a source of alienation as the car itself. The various voices that emerge from his chapter are spending at least as much time trying to reconcile themselves to the critical discourse on the car and the guilt they are expected to feel from their use of the car, as they are to their driving of the car. This is because for them even where the discourse achieves a much greater conscious-ness of externalities, these would still have to be brought within concerns that cannot be separated off as simply environmental issues. As I have argued elsewhere with respect to ethical consumption (Miller forthcoming), we see here a conflict between an ethics which is concerned with aggregate effects of personal action on the world at large and a morality that sees caring in terms of more immediate concerns such as one's partner and children. Thus the problem becomes one of whether we make a special car journey with adverse environ-mental effects because otherwise we feel we are exposing our children to discomfort or even danger. Without empathy the critique of the car comes across as the cold adversary rather than the warm friend of a humanity that defines itself as involved in an ordinary struggle between contradictory moral strategies.

It is women, in particular, who tend to be faced with these contradictions as yet another burden of the daily calculations of moral action.
Indeed the topic of gender is one of the most productive for making
explicit many complex issues of car culture. These contradictions are
illuminated by historical work on the topic, which shows how the clear
gender divisions in car use might be viewed as much as an unusual
foray by males into an otherwise female-dominated world of consumption as a struggle by females to prevent their exclusion from an
arena of consumption associated with male technological issues (see
Scharff 1991 and McShane 1994: 159–71) – the struggle for control of
the back seat complements any struggle over driving. What tends to
be missing is a more empathetic account of the centrality of car use in
women's lives. There is nothing equivalent to DeVault's (1991) sensitive
work on the juggling involved in the provision of meals that would
look at how a car journey must often also be a strategy about how to
fit a visit to the bank between picking up one child from school and
depositing another at his or her friend's (though see Rosenbloom 1992,
and Wachs 1992). Yet these are precisely the kinds of dilemma that
underlie the discussions reported by Maxwell.

For this reason it seems fitting to end the volume with a chapter by
Gertrude Stotz that takes the issue of gender as central, not in a narrow
form, but one in which gender is fully implicated in other issues such
as racism and the larger relationship of power between colonial culture
and the colonized. There is an intended symmetry here created by
starting the volume with Young and ending with Stotz. In order to
establish the manner in which this book acts as a breach with the
conventional literature Young's Chapter Two breaks apart a Western
discourse that would make us oppose the supposed authenticity of
Australian Aboriginal people to the supposed inauthenticity of the car.
The humanity of the car is first established in the degree to which it
clearly becomes an authentic instrument of Aboriginal culture. But once
that humanity is acknowledged then it is equally important that all
the faults and frailties of that humanity are also attributed to car culture,
which is what becomes evident in Stotz's Chapter Ten.

Chapters Ten and Two differ not only because they relate to two
quite different Aboriginal groups, not only because Two is dealing with
private vehicles which are much easier to assimilate than the communal
Toyota that is the subject of Ten, but also because the authors of the
two chapters intended to be complementary. While Young demonstrates in Two the integrity of the Aboriginal car, Stotz highlights in
Ten the contradiction that it brings out in a situation riven by conflicts

of racism, gender and, more generally, power. If anything the Toyota compounds the internal power conflicts of the Warlpiri with the asymmetries of gender relations assumed by the surrounding whites, such that the arrival of the car becomes a kind of Trojan horse containing gender conflict within its body.

Stotz in Chapter Ten then highlights and completes the intention of this volume to turn the study of cars into one of car cultures. This involves an initial acknowledgement of the humanity of a machine that has become integral to cultural life for so many people, an acknowledgement that includes rather than excludes all those contradictions that make an attempt to understand the car inevitably part of that larger struggle to understand our humanity. The volume thereby represents a third force in the literature on the car. This literature emerged with a history of events and persons broadened into a more general set of regional social histories. More recently there emerged a concern for externalities that is all those consequences of the car that had been removed from the frame of enquiry. With this volume the approach is to re-enter diverse historical and social contexts which problematize any assumptions that we know what the car is, except by developing a greater sense of what it has become and what cultural practices, values and moralities have become associated with it.

Obviously this volume makes no claims to be comprehensive; the regional coverage is evidently partial, as are the topics covered, with no chapter on car sales and purchase, for example. There is room for many related volumes on, for example, two-wheeled vehicles, or trucks in diverse regional and social settings. But it is the ambition of this volume to make such absences the more glaring. This can be only an initial account, and the various claims made in these chapters are thereby limited in turn. For example, I began writing this introduction with the clear ideal that if one could account for the massive discrepancy between the extensive treatment of other aspects of material culture such as food, clothes and houses and the paucity of any comparable studies of the car this might itself provide insights into the nature of car culture. I end defeated – I have very little idea why this discrepancy exists to the degree to which it does – but as with all the contributors to this volume at the very least we hope to generate a sense of what has thereby been missing.

References

Baran, P. and Sweezy, P. (1968), *Monopoly Capital*. Harmondsworth: Penguin.

Boge, S. (1995), 'The well-travelled yoghurt pot: lessons for new freight transport policies and regional production'. *World Transport Policy and Practice* 1:1, 7–11.

Brilliant, A. (1989), *The Great Car Craze: How Southern California collided with the automobile in the 1920's*. Santa Barbara: Woodbridge Press.

Callon, M. (1998), 'Introduction'. In M. Callon (Ed.), *The Laws of the Market*. Oxford: Blackwell.

Davis, R. (1992/3), *Death of the Streets*. Hawes: Leading Edge Press.

DeVault, M. (1991), *Feeding the Family*. Chicago: University of Chicago Press.

Dunn, J. (1998), *Driving Forces*. Washington: Brookings Institution Press.

Flink, J. (1975), *The Car Culture*. Cambridge, Mass.: MIT Press.

Freund, P. and Martin, G. (1993), *The Ecology of the Automobile*. Montreal: Black Box Books.

Gartman, D. (1994), *Auto Opium: a social history of automobile design*. London: Routledge.

Graves-Brown, P. (1997), 'From highway to superhighway: the sustainability, symbolism and situated practices of car cultures'. *Social Analysis* 41: 64–71.

Hebdige, D. (1981), 'Object as Image: the Italian scooter cycle'. *Block:* 5, 44–64.

Holtz Kay, J. (1997), *Asphalt Nation*. Berkeley: University of California Press.

Jennings, J. (Ed.) (1990), *Roadside American: the automobile in design and culture*. Ames: Iowa State University Press.

Johnston, J. (1997), *Driving America: your car, your government, your choice*. Washington: American Enterprise Institute Press.

Latour, B. (1993), *We Have Never Been Modern*. Hemel Hempstead: Harvester Wheatsheaf.

—— (1999), *Pandora's Hope: Essays on the Reality of Science Studies*. Cambridge, Mass.: Harvard University Press.

Löfgren, O. (1999), *On Holiday*. Berkeley: University of California Press.

Manning, F. (1974), 'Nicknames and number plates in the British West Indies'. *American Folklore* 87: 123–32.

Marsh, P. and Collett, P. (1986), *Driving Passion: the psychology of the car*. London: Jonathan Cape.

Mauss, M. (1966), *The Gift*. London: Cohen and West.

—— (1979), *Sociology and Pyschology*. London: Routledge and Kegan Paul.

McShane, C. (1994), *Down the Asphalt Path: The Automobile and the American City*. New York: Columbia University Press.

Miller, D. (1994), *Modernity: An ethnographic approach*. Oxford: Berg.

—— (forthcoming), *The Dialectics of Shopping*. Chicago. University of Chicago Press.

Moorehouse, H. (1991), *Driving Ambition: an analysis of the American hot rod enthusiasm*. Manchester: Manchester University Press.

Nadis, S. and MacKenzie, J. (1993), *Car Trouble*. Boston: Beacon Press.

O'Connell, S. (1998), *The Car in British Society: Class, Gender and Motoring 1896–1939*. Manchester: Manchester University Press.

O'Dell, T. (1997), *Culture Unbound: Americanization and Everyday Life in Sweden*. Lund: Nordic Academic Press.

Parker, I. (1999), *Traffic*. Granta 65: 9–31.

Perkin, H. (1976), *The Age of the Automobile*. London: Quartet Books.

Porter, R. (1999), *Economics at the Wheel: the costs of cars and drivers*. San Diego Academic Press.

Rajan, S. (1996), The *Enigma of Automobility*. Pittsburgh: University of Pittsburgh Press.

Rosenbloom, S. (1992), 'Why working families need a car'. In M. Wachs and M. Crawford, (Eds.), *The Car and the City*. Ann Arbor: University of Michigan Press, pp.39–56.

Sachs, W. (1984), *For Love of the Automobile: Looking back into the History of our Desires*. Berkeley: University of California Press.

Sawyer, M. (1999), *Park and Ride*. London: Little Brown and Company.

Scharff. V. (1991), *Taking the Wheel: Women and the coming of the motor age*. New York: The Free Press.

Schiffer, M. (1994), *Taking Charge*. Washington: Smithsonian Institution Press.

Slater, D. (1997), *Consumer Culture and Modernity*. Cambridge: Polity Press.

Thrift, N. (1996), *Spatial Formations*. London: Sage.

Urry, J. (2000), *Sociology Beyond Societies*. London: Routledge.

Wachs, M. (1992), 'Men, women and urban travel'. In M. Wachs and M. Crawford (Eds), *The Car and the City*. Ann Arbor: University of Michigan Press.

Wachs, M. and Crawford, M. (Eds) (1992), *The Car and the City*. Ann Arbor: University of Michigan Press.

Weiss, B. (1996), *The Making and the Unmaking of the Haya Lived World*. Durham: Duke University Press.

Whitelegg, J. (1997), *Critical Mass*. London Pluto Press.

Whitelock, F. (1971), *Death on the Road: A study in social violence*. London: Tavistock.

Wolf, W. (1996), *Car Mania*. London: Pluto Press.

Note

1. Thanks to Elizabeth Shove for this observation.

The Life and Death of Cars: Private Vehicles on the Pitjantjatjara Lands, South Australia

Diana Young

Long time ago people had no car, they had to walk. They walked miles to get kangaroo meat and bush tucker... Today we have cars to go anywhere we want to or to go hunting.

Bertha Nakamarra (1991: 18)

This chapter is about some aspects of the life and death of cars on the freehold Pitjantjatjara Lands in the Western Desert, South Australia. I will argue that Aṉangu[1] use cars as social bodies. The driver of a car must imitate the complex choreography of spatial etiquette of a person on foot. The car is therefore almost a prosthesis of the persons inside and its carefully constructed uses express much about spatial practices in contemporary Aṉangu culture. Similarly abandoned cars, which rot extremely slowly in the dry desert, are used as spatial markers in country but are also akin to bodies.

Fieldnotes 1997. *On the main road the red car comes towards us lights flashing and arms waving from every window telling us to stop. We draw up, the drivers able to touch. Kunmanara gets out and D passes her baby grandson through the window to her. N and M get out of the red car into the white one with crowbars and a blanket. We drive back to the homeland to fetch the shovels, then off to the west, 'same place', for honey ants.*

35

Figure 2.1 Abandoned Dodge, resprayed bright yellow, missing its bonnet, engine, wheels and lights.

Of all Western consumer goods, motor vehicles seem the most important to Anangu. Life without cars in remote bush communities is considered impossible, despite the fact that two generations ago most people walked long distances every day. No one now would consider a hunting expedition without a car. There are fewer emu and kangaroo and one has to drive further away from settlements to find them.

Relatives always need a lift to the shop, to work, to fetch the mail and pay. Twice a week when the mail plane comes in from Alice Springs carrying wages, the centre of the community is thronged with cars each bursting with people. No one except some very old people born in the bush wants to walk anywhere if it can be helped. The younger you are the less likely you will want to walk.

Both Pitjantjatjara and Yankuntyjatjara are spread over a wide area of South Australia and the Northern Territory. There is constant travel for business and other social reasons (such as football and funerals) between Ernabella, Fregon, Amata and Mimili. The Aboriginal community of Yalata for example on the south coast of South Australia, is predominantly Pitjantjatjara as are Areyonga and Docker River in the Northern Territory. Many Anangu live in Adelaide and Port Augusta in

South Australia and in Alice Springs and Cooper Pedy. With family networks extending over such distances, individuals may choose to stay with relatives hundreds or even thousands of kilometres away, for weeks or months at a time. There are always cars going between these places and people decide to go when they wish often on the spur of the moment. This simultaneous expansion of horizons – 30 years ago few people in Ernabella had been as far as Alice Springs – and contraction of time-space in terms of distance is similar to that experienced in Europe with the advent of rail travel (Shivelbusch 1986).

Car ownership is important in this interplay between individual autonomy and the continuous dynamism of social relationships, analysed eloquently by Myers working with the Pintupi (Myers 1991a,b). Access to motor vehicles is indispensable for fulfilling social obligations and enhancing political status. At the same time, mobility is a means to avoid social conflict, often cited by bush people in their desire to live and move in smaller family units on homelands more similar to those of the past/*iriti* before white people came (Hamilton, A. 1987: 48).

'Two of the most striking features of present day aboriginal religious life are its high mobility and the far reaching communications network' (Kolig quoted in Hamilton, A. 1987: 48). Women and men travel huge distances for ceremonies, or business, as it is now termed by Western Desert people. Without access to cars tjukurpa[2] sites cannot be visited and secret single-sex ceremony cannot be produced without anxiety. This is more of a problem for women than for men since it is men who generally control access to cars, both private and community vehicles (see Chapter Ten).

But it is not just the matter of needing a car for obtaining firewood and food from the store or the social necessity of attending funerals, openings, football and *inma*,[3] but the emotional compulsion to be mobile, to be moving through country especially one's own. As women often told me, just being in the bush makes you feel better.

Acquiring Cars

The relationship between people and cars is by no means an unproblematic one and is further complicated by a vehicle having community status (Stotz 1993). In this chapter I will concentrate on private cars.

Anangu in the Western Desert acquire motor vehicles in the same way as the non-Aboriginal population, through cash purchases from private sales and second-hand dealers, in Alice Springs, the roadhouses at Kulgera and Marla Bore as well as from further afield – Adelaide and

Port Augusta. Sometimes non-Aboriginal staff in Ernabella bring cars up to the Lands to sell. While a minority of high-earning families buy new vehicles – almost always four-wheel-drive land cruisers – second-hand vehicles are more usual. Cars are sometimes 'swapped' either for other cars or for music systems and televisions. Several of the cases I recorded were swaps between siblings, but one was with a white man when a newly acquired minibus got stuck in deep creek sand on station land and was abandoned there to be retrieved by the white man in exchange for his car.

The cars that Aboriginal people in the Western Desert buy are typically in the last stages of their viable life. Indeed new cars are often shunned as unpredictable, 'might be bad luck', perhaps because they are unsocialized – that is, they have no accumulated identity (Stotz 1993: 193).[4] In Ernabella the annual receipt of tourist gate monies from Ayers Rock by those who claim traditional ownership of sites inside the National Park causes an efflorescence of vehicle purchasing. Even so, most of these cars will not last long. Some sedans (saloons) and four-wheel-drive vehicles as well, last only a few months before irretrievable breakdown occurs on the punishing dirt roadways.

Both Annette Hamilton working in the Western Desert with the Yankuntyjatjara and Fred Myers with the Pintupi, have documented how consumer objects are acquired and disposed of with ease and assert that cars are no different in this respect (Hamilton 1979: 111; Myers 1991a: 73). Myers contrasts the inalienability of the land (since enduring social identity is formed by a person's relationship with country) with the way 'most objects enter into exchange' and this includes cars. Whatever was obtained once can be obtained again (Hamilton ibid). In the east of the Pitjantjatjara Lands, much of what is acquired in the way of mass-produced Western goods – clothing, blankets, car-tyre rubber, toys, household goods and packaging – is discarded, apparently randomly, onto the ground. The soft sand gradually swallows these things and they submerge and disappear. There is also a sense in which letting Western things go into the ancestral earth ensures their reproduction. Car's bodies however do not enter the sand and rust slowly in the dry climate. Even burnt-out cars remain as skeletons. Although then it is literally out of circulation, by its continuing physical presence in the land, a car never really dies.

Anangu say that they cannot 'look after' country without access to cars, notably Toyotas (four-wheel-drive vehicles). In my experience 'looking after country' now often implies residence and may be a basis for claims to that country, and this is corroborated by Payne who worked in Ernabella (Payne 1984). Such residence on a homeland is impossible without a car. 'Looking-after' is also achieved by performing ceremonies. Cars, although desirable, are not 'looked-after' in the way that people or country are ideally stated to be.[5] However Anangu, I suggest, perceive cars as having a special relationship to country unlike that of other things, especially other mass-produced commodities, because they are literally the vehicle for social events and the places and journeys which they enabled. I will examine this notion more thoroughly below. Cars in the Western Desert are a better example of things having a social life than even Kopytoff might have imagined (Kopytoff 1986: 67).

The aspect of a vehicle Anangu seem to consider as a constant throughout its biography resides in its materiality, notably its colour. Thus Hamilton records that one of the first vehicles owned in nearby Mimili in 1970 was known as 'the Red Truck' (Hamilton 1979: 110). My old Toyota troop-carrier is recalled as 'your white car', although as I will show it had other social identities. Yami Lester, a Yankuntyjatjara man, remembers the red Ernabella truck in about 1954 (Lester 1993: 40). Even when abandoned, sold or swapped, or carrying contrasting colour panels taken from dead vehicles, the car is still 'that red one'/ 'palatja rituwana' or 'that yellow one', and colour is a common way of identifying many other things as well in everyday speech and circum-stances. The set of tracks left by a vehicle's tyres, the walka/design on them, also lead to the identification of a car's passage and much can be inferred from them about the likely occupants and when and where they went. Anangu also recognize individual vehicles by their engine sound although on the whole people seem to do this less accurately than with tracks.

When travelling outside their own community cars are identified with their home/ngura. So people might identify a vehicle driving round Ernabella as a Fregon car. When I drove to a mining town several hundred kilometres south, my car was identified and hailed by Anangu living there as 'Ernabella car'.

Private cars are also verbally identified with individuals, though this seemed to happen when a vehicle had belonged to that person for a longer time than average. Such a car will then be identified as such and such a person's when being driven by others, with the owner absent. Often relatives help pay off car loans but the woman or man in whose name the loan was taken out remains the controller of the vehicle. What happens to a car at the owner's or habitual user's death is considered below.

Inside and Outside Cars

Cars, like most other things, have many uses just as a car journey has no single *raison d'être* even if only one is stated. Among many reasons for cars being so desirable is their quality of being a convenient mobile camp and ready-made architectural element. My own car, because of its size and status among a group of women, had by the end of the fieldwork become '*minymaku*'/women's Toyota. It was used to store all sorts of things, some temporarily, others (including a rifle) for several weeks. Clothes, women's ceremonial 'tools,' bags, and boxes of groceries, crow bars, supplies of bullets and innumerable tobacco tins of *minkulpa*/ bush tobacco all passed through it. Often I had no idea what was stored inside, especially the business 'tools' which seemed to be safe and out of sight there at a time of frantic ceremonial undertakings. I was regularly approached by people searching for missing bags, medicines or shovels which had at some time passed through the vehicle. Glove compartments are used to store small or comparatively precious items, cash, tobacco, a bible or hymn book. New passengers in a car thus always seem to rifle through the compartment first and may appropriate something they want. Nungalka, finding her crowbar missing from my car, took a hand-crocheted beanie (hat) instead, saying to me in Pitjantjatjara, 'someone takes my crowbar, I'll take their hat'.

Although living in old car wrecks was sometimes practised in central Australia this is not the case around Ernabella. But a live car provides all the useful places for stowing equipment and belongings away from the demands of others afforded by the *wiltja*. *Wiltja*, meaning shade, are circular or semi-circular in plan, made with live-leafed branches or with dead ones. The timber has one end embedded in the ground and the top parts are pulled together and interwoven with smaller forked branches.[6] The frame is covered in tarpaulins or in the past, spinifex grass. A doorway is formed by omitting several of the uprights. 'During the day, the *wiltja* may be used as a "shade", to give protection from the sun's heat, for privacy, or as a place for social retreat associated with avoidance behaviour. The *wiltja* is also used as a place for storing personal ritual objects to be kept out of sight' (Hamilton, P. 1972: 7–8). Meat and clothes are stored on the *wiltja* roof, safe from the depredations of camp dogs.

Private motor vehicles are used in all these ways too (although meat stored on a car roof is often rendered accessible to agile dogs via the handy platform of the bonnet). Except in cold or wet weather, Anangu do not often sleep inside *wiltja* at night (or indeed inside cars) but outside in the *yuu*/windbreak area where the fire is made. It follows then that on homelands, houses are often used for storage while people live either in *yuu* structures that they have attached to the verandahs or, away from the house, but within the fenced compound, in a *yuu*. The *yuu*, an adjunct to the *wiltja*, is made from short leafy branches loosely interwoven with others. In communities it might be made from bits of corrugated iron. Bigger timber structures are hung with cloth or blankets for visual privacy.

Camps are always in a state of flux, their material aspects adapted due to prevailing wind or rain, the arrival or departure of visiting relatives and the current state of social relationships within the camp (Hamilton, A. 1979: 23–4). Although many people spend much of the year living in houses in larger communities, there is constant away camping during ceremonial business and, sorry camps,[7] by old people and by those who live on homelands and camp both there and on other occasions in the community. While *wiltja* and *yuu* may be swiftly though laboriously altered, a live car can be used to alter the living spaces simply by someone turning on the engine and moving it a few feet, changing the angle, etc. When first asked to move a car in camp in this way I had no idea what I was supposed to be doing much to the hilarity of my instructors. Camps planned around cars looked, to me, bleakly exposed during the day when the occupants are out driving in the vehicle.

When neighbours on the homelands purchased a full-size single-decker bus, which they said was 'like a caravan',[8] it did seem that in many ways this was their ideal mobile *wiltja*. It quickly filled up with their possessions, cloths, tape recorders, shovels, etc. and the whole extended family of three generations could fit into it albeit travelling at a stately pace. Usually they owned three or four cars between them. The bus engine failed within a month and the idea was abandoned. However, many families own minibuses although their low undercarriage means that they are often stranded in soft creek sand where their owners take them to transport roots of the red river gum for carving. Other people will always pass in other cars and get them out even without the benefit of four-wheel-drive vehicles, with a mixture of whatever towing equipment, jacks and shovels and much manual labour.

At women-only ceremonies, always held in the bush, parked cars screen the singers and dancers from on-coming vehicles. A second car is parked slightly set back from the first and so on, offering information to later arrivals about which group arrived when. This in turn is related to individuals' and groups' rights to sites and to the enaction of the *inma*. At a boys' initiation camp cars were parked in a dense group, making – to those outside – an invisible ring at their centre. Car headlights are used during nocturnal ceremonies to illuminate the dancers, augmenting large bonfires replenished continually for the same purpose. Toyotas are preferred for this possibly because–the lights are more powerful and the battery is larger and can stand the strain. Toyotas seem anthropomorphised in these circumstances their slow movements apparently self-sufficient and choreographed.

At public meetings or church *inma* drivers group their cars in a rough semi-circle facing the action, sometimes banked two or three vehicles deep. Although Anangu usually sit on the ground, during bad weather at meetings or football matches people might sit on or inside their cars. At a sports carnival held during a dust storm, women moved from car to car, from group to group, climbing into each car in turn, just as they would do in camp between *wiltja* or in the community between houses. Cars have the added advantage of built-in radios and tape decks, a facility that young women particularly are addicted to. Aboriginal Police Aid vehicles which are equipped with loudspeakers are used to broadcast community information as they drive around as well as, on occasion, country and western music.

Men in particular (for it is men who are usually the 'bosses' of cars) may retreat alone to their cars for a smoke or to listen to music on tape or the local 5NPY radio.[9] Being able to do this is an expression of power. I laboured with three women to prepare and cook the game one of them had shot on a hunt while their (and by then, my) male relative sat cocooned in the car, smoking and listening to a music tape. Sometimes men simply sit inside their cars amidst a family camp. In women-only ritual gatherings women often withdraw to the interior of the car they have come in, to eat or drink. If you do not want to be asked to share your food, cigarettes or *mingkulpa*/tobacco, it is advantageous to conceal your consumption, for to turn down requests flat without an excuse is to deny your social relationship with the other person (see Myers 1991a: 56 on the Pintupi).

Children, from babies to teenagers, are left to sleep inside open cars in the day on these and bush tucker trips, where they are safe from snakes. In the event of an argument breaking out those who wish to avoid confrontation retreat to the car and move off a little way in it, if possible. In all these situations the car is used as a liminal space, separating those inside from social demands, avoidance relationships or aggression in a way similar to the *wiltja* (P. Hamilton 1972: 7–8). But the car's capacity for motion gives more scope for such retreats; thus women at boys' initiation camp are able to go back to their own camp in cars and avoid meeting male relatives, which they would have had to take much trouble not to do on foot. Most of all a car universally gives one the ability to leave – quickly. At the sorry camp of an important man in a neighbouring community of Ernabella, a dispute broke out among the mourners as to where the deceased should be buried. The family I was with made hasty preparations to leave and we went instead to a party at Ayers Rock but returned to the sorry camp in the small hours of the next night when all the trouble – and the funeral – had finished.

A car, then, can be used both as a *wiltja* and as a super-*yuu* providing visual and some auditory privacy, often as an addition to other camp structures. On one night camps cars are simply angled appropriately in relation to other camps or to the prevailing weather, or in relation to features of the surrounding country. A fast variant on a wiltja in the middle of a suddenly rainy night is achieved by parking a car to form a wind-break and attaching a tarpaulin to its roof to make a lean-to shelter. Cars parked for shelter at night are moved at sunrise to let the sun's warmth fall on the occupants of the camp.

Going Along

As they travel along dirt roads[10] cars are transformed, rapidly modified both by the bush through which they travel and by their owners, acquiring contrasting panels, odd wheels, fabric instead of window-glazing and, inside, an accumulation of rubbish. The further they travel the more modifications occur. The orange desert dust makes red cars almost invisible in the distance and white ones especially become gradually transformed until their lower panels are the same colour as the ground. Several Anamatyere artists, of whom Mavis Holmes Petyarr is one, working from Utopia in the Northern Territory made paintings in the late 1980s using old car panels as support. The head of a person is painted on the window as though going along inside the car, and red and green country – rocks, trees, earth – is depicted on the metal paintwork part of the door. I imagine that the artist was showing how the travel of the car through that country had transformed both the vehicle and those inside it through access to ancestral power – that is, through access to country.

Women are constantly seeking time and opportunities to go on bush trips looking for honey ants/*tjala* and witchetty grubs/*maku* and during August and September, goanna/lizard as well as hunting for larger game of which the commonest is kangaroo. There are regular sorties to obtain the roots of the red river gum for carving and *mingkulpa*/wild tobacco is gathered when ever possible, as it is addictive. One of the most pressing reasons to gain access to cars for men and women alike is for the production of religious ceremonies.

It is through driving around the country – tens of thousands of kilometres – mostly with women, that I learnt something about their lives and their country and how to drive appropriately in both communities and the bush. I was in effect socialized by the fortuitous possession of a large, ancient personnel carrier.

Passing near a women's site or the tracks connecting sites, women sing the inma connected with the place. Sometimes, they sing Pitjantjatjara hymns from a Christian *inma* the night before.

A group of persons in a car generally stay together and camp together when arriving at a large gathering. Thus, picking up a group of people to travel to such an event may last hours. Not only is everyone rounding up necessary material items – blankets and mattresses, food, clothes, crowbars, etc – but they are also negotiating the composition of the group in the car and this will depend upon kin relationships and political manoeuvrings of the moment in the constant flux and renegotiation of social bonds. The car then does provide 'a temporally extended objectification of shared identity' (Myers 1991a: 55). The pick-up involves serial circuits of the community, importuning others, searching for children, missing equipment – billy-cans, water containers, etc. – and acquiring food and drink supplies, as often from kin as from the store. Tiny cars and petrol tankers too are one of the most popular toys for both girls and boys and are often bought for them at a store on the journey. Toy cars are soon discarded, like their bigger counterparts rapidly losing wheels, but unlike them, disappearing into the red sand.

Inside the car, this arrangement of people hierarchically and spatially in relation to the front of the car is ideally the same as in camp in relation to the fire when sleeping. The most senior person in kin terms sleeps next to the fire with others ranked behind. The most senior passenger in relation to the driver (spouse if present) sits in the front passenger seat. Behind, other passengers sit ranked from front to back according to their relationship with the driver. Often two competing powerful women jam into the passenger seat together.

Through the care taken by the driver and the instructions shouted by others, the car must be made to navigate space as though it was a person. Vehicles must be made to conform to strict spatial rules, aligning the car precisely under the shade of a tree, or in the sun on a cold day for the comfort of those it carries. As a social body the car must approach persons with the same attention to proximal etiquette as a person on foot. For example if approaching a group seated on the ground, one should cough or call from a distance and wait to be invited over. Similarly, unless a close relation one does not drive into the fenced front yard of a house, but waits in the roadway and shouts or hoots until the occupant appears and comes out to greet you.

One does not make tracks on other peoples' land, which includes their domestic space of the moment, unless invited to do so. When visiting another's homeland, an event unlikely to take place without a vehicle, Anangu wait inside the car until greeted even if the place is the residence of close family. When dropping someone off with her bags and bedding after a camp or with other produce gathered from the bush on a day trip, goods are unloaded by the passenger outside the house and dragged inside the fence by the occupants of the house later. The same curtilage of private space seems to exist around houses now as in the past around *wiltja* (P. Hamilton 1972: 8).

Anangu make tracks with the car on the country which are intelligible in the same way as those made travelling on foot[11] and Anangu drive cars across the bush as though on foot when in pursuit of kangaroo or emu. Women have used cars to circle likely-looking goanna burrows when hunting (to ascertain if the lizard has made recent tracks), making the car turn tight circles to view the burrow from all angles just as they do on foot. Women will drive right up to bushes to harvest leaves or fruit. Similarly, car tracks go as far as is practically possible to open sacred sites – that is, sites that men women and children may all visit.[12] Older experienced women drove the heavy Toyota across bush land with great skill, avoiding hidden *punu*/logs and tree stumps, *piti*/rabbit warrens, *puli*/rocks. Driving after powerful *inma* must be done with care. Similarly at ceremonies associated with mortuary rites, the ground is disturbed as little as possible by cars on their way to and from the grave, by driving slowly in single file, each car keeping closely to the tracks of the car in front.

Anangu, then, can track a motor vehicle in the same way as they can track a person or animal, and in many ways the marks a vehicle's passage makes on the land are conceived as the gestures of the driver. Hunting roads made by driving across bushland are named as the driver's road. One who makes the road on the outward journey should drive back along it. (The ability to see tyre marks as faint depressions in springy grassland, for example, was one I did not share with my passengers: 'Kuru wiya'/no eyes!) . 'Para-pitja' translated by Anangu as 'go round' was a habitual instruction to me whether in a community or in the bush. My attempts at reversing or three-point turns were classed as 'wrong way'. Anangu point out tyre tracks revealing a vehicle's passage as it reverses as whitefella's or tourist's tracks. All around communities on the Lands are 'roundabouts' where cars entering a confined residential area habitually turn, perhaps round a rock or a telegraph pole. Furthermore turning is ideally done in an anticlockwise direction. Roads with many bends are described as *kalikali*/crooked, bent (*kali*=boomerang) or *'liru ruta'*/snake road because the tracks the car leaves and the way it moves are like those of a snake.

If single cars out bush are driven gesturally rather than with the intricacies of camp and community social space in mind, gatherings of many vehicles are choreographed when arriving en masse at large camps, especially where outsiders are present. Ernabella cars wait for one another and enter such gatherings in stately single file, keeping close together and creating a dramatic and moving impact of power and solidarity.

'Nganampa ruta'/our (collective) road, *'minymaku ruta'*/women's road, the history of roads was often told *as we went along*. This is the old *Finkeku ruta* (possessive, belonging to Finke), this is where we used to come on donkeys, this is the old cattle road, this is the old Amata road. These things are generally only spoken of as they are passed. Older people especially use hand signs to commentate on what is passing, but no one mentions abandoned cars.

Roads are a major form of human mark-making on the country, all witnessed or made by the last two or three generations. The narrow tracks – *iwara* – through the ranges are no longer trodden and within another generation may become lost to memory.

Anangu know dirt roads, their potholes, bends and rocks as well as they know the surrounding country and shout out warnings to the driver accordingly *'purkara/*careful slower'. Women remark on old honey ant pits that they dug in the past, another habitual form of human mark-making on country, and events that happened on the road caused by the interaction of car with country: 'We hit that big bump there and we was all crying (with fright)', or point out where, out hunting, Kunmanara drove into the wire fence while looking the other way at *malu*/kangaroo.

Navigation

All the driving I have done with Anangu was in country and communities known to at least some of the occupants of the car. In these conditions Anangu seem to remain aware of the position of the cardinal points all the time, inside the car or out of it (although people often check the position of the sun) and small children in remote communities are knowledgeable early about their orientation (see Lewis 1976). Asking someone where they are from, for example, elicits silent pointing towards the direction of their place.

The ability of Aboriginal people to navigate in the bush is legendary and one of the few things, along with tracking, which non-Aboriginals admired them for. Anangu are well aware of the vulnerability of whitefellas left alone in the bush who have no idea where their car is after taking 100 paces away from it in Mulga scrub. When walking away from the car, hunting for signs of honey ant among Mulga trees, women seem to note intimate details of the land, a fallen tree, animal tracks, things which stand out as different or contrasting in colour, but in the car in open country more distant landmarks are used too. Older women's knowledge of the local *tjukurpa* places, created where ancestors pausing to sleep, eat, defecate, etc. as they moved through the country, is embedded in their navigational skills as such places are often also prominent landmarks.

Left/*tjampu* and right/*waku* were not used as direction or orientation words in the past (Goddard 1994). Anangu use these terms when directing non-Aboriginals but among themselves use cardinal points. Possibly left and right terms were used only for things in direct association with the body (Wassman 1994), as whether someone is right- or left-handed is always known by others. At a night camp when it was critical that women leave fast after a ceremony, I was told to park the car facing north. As they were preoccupied by the potential danger of the situation, their usual allowance for my ignorance of the cardinal points went astray. Eventually I asked 'You mean towards Ernabella?' 'Yes, yes,' they said, aware perhaps that both were equally impossible for me but were habitual to them.

If, then, in communities cars are moved to respond to the shifting dynamic of social space as well as the time of day and the weather, navigation in a lone car driven across open country is in response to the absolute cardinal points and to landmarks (Lewis 1976) and to the mental imagery of the navigators. Nura R. would say, tapping her forehead, '*kulini na* – thinking', meaning 'I remember, I know'. Says one of Lewis's informants, the Pintupi Jeffrey Tjangala, 'I keep these (north south east and west as a cross) inside my head.' (ibid.: 264) I do not wish to confuse language (words and sign language), concerned with space, with spatial cognition (Wassman 1994). However, in

Pitjantjatjara/Yankuntyjatjara, *kutu* is a case ending used to indicate orientation towards, usually in association with movement (Goddard 1994). When travelling, the question 'where are we going (towards?)' will be answered either towards place X or towards a feature – sandhill/ rock, etc. – or it could be answered with cardinal directions. Most likely the answer will be gestural, either with hand signals or with the mouth. An established road, that is a socialized route, is named for the place it goes towards. Anangu might say that the main road towards Ernabella is the Ernabella road but travelling the same road towards Amata, a community to the west, the road becomes the Amata road.

'Short cut' can mean either a wavy circuitous route often to avoid other people or places or a direct straight line back to a base point, for example after the a long zigzagging pursuit of kangaroo. When asked directions, Lewis found that the Western Desert men with whom he worked visualised themselves at some often distant point of reference from which the directions are given (often a sacred site) and gestured as though they were there. Ernabella men and women who have long experience of non-Aboriginals' route-finding, draw maps for white people away from their bodies, so that the bottom of a sheet of paper nearest to their own body is our present position and the route straight up the page leads to the destination. Recognizable features along the way which cross or lie adjacent to the route are listed as *karu*/creeks, *puli*/rock or hills and to a lesser extent, *punu*/trees.

Anangu move around in a car as though it were an extension of the body because, I suggest, their sense of space is not *only* an egocentric one such as Kant hypothesized as universal and westerners regard as 'natural' (Wassman 1994). Anangu move the car not in response to some abstract spatial rules, but in a way that, in the bush especially, is in relation to country and articulates a spiritual and emotional tie to the country and to the navigational skills of hunter-gatherers travelling on foot. Of course, when on the main public roads, such as the Stuart Highway which runs from Adelaide through Alice to Darwin, and in towns, Anangu are perfectly able to drive according to the highway code[13], although many of the women who are such good bush drivers are not so confident in towns away from the perceived safety of the Pitjantjatjara freehold Lands. This is partly due to unease at the number of white men also using the roads.

Dead and Declining Cars

Feb 8th 1998 *Spent last week taxi-ing Vic Well mob about for spanners, petrol, etc. White Ute, white sedan and coffee-coloured mini bus all sick apparently. Clem removed petrol pump from white ute stuck at Ten Well and took it to Vic Well to fix into another car, pale yellow/coffee. He had to go into Ernabella to buy a set of spanners, although he tried to borrow first from N. on another homeland. No sign of white sedan.*

The battery is one of the parts of the car which receives attention in case of breakdown (see Chapter Ten). Batteries are regularly transported into Ernabella garage for recharging and jump starts with jump leads or push starts, by ramming the car, preferably with a large Toyota, are a way of life. Batteries, wheels and even engines are often transferred between cars in an extended family. Young men in particular, having few employment opportunites, may spend weeks bent over car engines or combining two or more cars. Since most cars on the Lands are still of a few preferred makes only, this is easy to do, but great ingenuity and imagination is exhibited in order to keep at least one car of the ailing group capable of going along. Cars are still usable without brakes, lights, reversing capacity, petrol gauges or windows, to give but a few examples. These kinds of car, always in need of mechanical attention before they will leap into motion, are classed by Anangu as 'rubbish cars'. Nearly all cars eventually become rubbish cars. Such vehicles are habitually stranded on the roadside nose-to-nose with another, jump leads strung between open bonnets.

Figure 2.2 Car bonnet re-used as a signpost.

A car that no longer goes, as all machinery, is *pikajara*/having sickness, or *pika pulka*/very sick, or *ilunyi* (*ilura pupanyi* lit. crouching dead/ unconscious (Goddard 1994; Hamilton, A. 1980: 110), the same descriptions used of persons. The euphemism for a person's death *wiyaringanyu* finished, is now more often applied to both vehicles and people than is *ilunyi*. Anangu class game such as kangaroo, as *wiyaringanyu* when the animal is unable to move along, but is not necessarily dead. In this chapter I have used the term 'live' cars for those that go along, but this is my construct and not one I have ever heard Anangu actually voice. However, there is evidence to suggest that the engine and the battery of a vehicle are regarded as having a life force. When engines[14] are swapped between car bodies the carcass without an engine is *una* which Anangu translate into English as 'useless'. Coming silently to rest as all the car's power cut out late at night on a potholed back road, and trying push starts to no avail, my women companions, muttering about my ignorance of car engines, insisted that the battery was to blame and gathering around the car bonnet prayed to Jesus.

Once a car is abandoned the wheels almost immediately disappear. These are always in demand not least because the pursuit of kangaroos across the bush often results in the complete loss of tyres. Cars travel short distances on bare wheels if need be. The next items to be removed are the battery and the engine. The body panels, the bonnet, boot and doors disappear too. Car parts are used to make other things, especially women's shallow digging dishes, *wira* out of hub caps. A high degree of ingenuity and improvization is employed in reusing unlikely found objects as well as parts from dead cars to continue the life of another vehicle. Car bonnets are regularly employed as support for written sign posts not only because they are large handy sheets of durable material but for the equally functional reason that they are the front of the car. The bonnet is the *mulya*/nose of a car and arrives before the rest of the vehicle.

Like live cars, dead ones and spare parts have many potential uses. Intact dead cars marooned outside houses are sometimes used as retreats by solitary young men. Anyone can take parts from abandoned cars if they are on the road but wrecks on or near homelands are the property of those who live there and/or the nguraritja/Traditional Owner if these are not the same person. Men visiting another's homeland might take the opportunity to peruse the dead cars inside the compound. Car dumps in communities are also 'public'. Several of the car dumps in Ernabella were near men's sacred areas. I know this because there were notices saying that they were sacred areas and out of bounds, in English.

Perhaps this dumping in an area inaccessible except to initiated men was a statement of rights over the car wrecks too. Cars that are first abandoned where they terminally breakdown, at the side of the road, gradually get pushed off it, sometimes upended or overturned as various parts are stripped. Many end up far away, but still visible from, the road. Dead cars often finish up under trees, or upended or on their roof.

The relationship between topography, dead cars and dead people is, I suggest, one that Anangu make (see Stotz 1993: 175 on the Warlpiri). Until the early 1970s Anangu buried their dead at or near where the death took place. The burial party was small and, after the decay of the body, the pit was covered and left unmarked. But the place was remembered. Since the early 1970s, Anangu have decided to bury the dead in Christian cemeteries. Each community has such an area, and homelands too. These graves are decorated and marked with headstones and crosses.

Among the Pintupi, says Myers, goods (including cars) are not allowed to carry the deceased person's identity forward in time (1991a: 71). Contact with deceased persons' things that they habitually used and the places they frequented makes the living sad – 'sorry' – and may attract the dangerous spirit of the deceased. These things are therefore destroyed or given away or, in the case of houses, special rules and rituals are applied to their cleansing. In the past when someone died the whole camp was torched. But for Anangu even in the 1990s the accumulated car wrecks are still present and so to an extent carry the identities of deceased drivers forward, reminding those who knew them of all the places and events where and when they drove those cars.

A homeland often has a grave to one side of the house and a car dump on the other. On many occasions abandoned homelands were pointed out: 'There is my uncle's grave (ngaltujara/poor thing)', while near the grave several abandoned cars went unremarked upon. Visiting her husband's decorated grave on their now abandoned homeland, his wife turned away from the pile of old cars he had once driven with a gesture of sadness and disgust. The dead cars seemed more of an affront than the house or the land they had camped on. The cars had connected her not only to him (she could not drive) but to all the places on all the days they had travelled the country together.

While bodies are confined to burial in restricted areas, cars maybe dumped in the bush. Thus when, for one reason or another, the motor car belonging to the deceased is not given away to kin in another community, but abandoned, the inverse of the traditional burial system occurs. Hunting goanna (lizard) in thick scrub some way from Ernabella, we came across an intact, newly abandoned vehicle. My companion immediately turned away from it and walked rapidly in the opposite direction instructing me to do the same, checking over her shoulder for mamu/harmful spirits. It was the car of a recently deceased man whose body was by then in the cemetery at Ernabella. This was not a remote spot but one frequently used for hunting. Both he and his car may cease to go around but whereas his body is confined to the cemetery, where the grave is labelled and decorated, the car is out in the bush, marking another spot that for those who knew him will make them 'sorry' and will be associated with his memory and where he went, and with whom, in that car. I found nearly as many taboos surrounding discussion of abandoned cars as around that of dead persons, whose name must never be mentioned, perhaps because the two were linked.

Conclusion

It was not until 1961 that a Pitjantjatjara person owned a motor vehicle, although from the 1930s onwards cars driven by Europeans would have been seen in the area (Edwards 1994: 148). In Ernabella children's drawings from 1940 onwards, cars were one of the first western things, along with windmills (for bore water) and horses, to appear. In the earlier drawings, cars are shown with totemic animals, emu, perentie lizards, kangaroo and elements of country, rocks, trees waterholes. In the collections of Ernabella schoolchildren's drawing that remain from the 1940s, '50s and '60s[15] cars and trucks appear frequently while camels, which were at this time the favoured mode of transport for many desert Aboriginal people, figure far less. Those children are the car owners of today. There was and is obviously something magical, enchanting about the car.

Why then are cars so desirable and so eagerly appropriated from among the plethora of western goods on offer? The desire for mobility by previously hunter-gathering peoples has been well documented and Annette Hamilton has stressed the desire for mobility for the maintenance of ritual connections in her work with Yankuntyjatjara people. Myers emphasizes the objectification of shared identity residing in the car as well as the need for autonomy. I have suggested that cars mediate, not only, the constant dynamic of social relations but also, crucially, the strong emotional relationship of people with country. Rather than rendering this relationship more distant or 'inauthentic', the car can reinforce Anangu's spiritual connection with the land. This connection with country then is not just one of identity but of a wider sense of well-being, of being in the land, literally in the case of digging honey ant pits. Yet travelling sometimes all day inside a car hunting for game seems to be just as satisfying emotionally as physical contact with the red earth, especially if *malu*/kangaroo is obtained.

I have also dwelt on the materiality of cars which helps rather than hinders this act of mediation. Work with hunter-gatherers has often stressed the simplicity of their material culture, and among Central Australian Aboriginal groups the Pitjantjatjara were particularly singled out in this respect (Tindale 1972). A woman's digging stick or a man's spear-thrower was used in many ways, including symbolically, where its material form evokes analogies to other things. That the car is a complex thing deployed in complex ways is evident even in the way it falls apart, is deconstructed and recycled. The materiality of the car also offers the possibility of doing certain things better – the production of religious gatherings for example – with more efficacy than before the car's existence. While it creates certain serious problems, petrol-sniffing for example, which I have not had space to dwell on here, the car also resolves certain other conflicts, as in the provision of a new kind of privacy within the very open nature of Aboriginal camp life.

Figure 2.3 Abandoned car seen from a main road.

I suggest that the capacity of the car to be an agent of transformation, in a camp say, as well as to become transformed itself through its journeys across country links it to the strong concept of ancestral transformations in Anangu thought. In short, Western things which embody this capacity – clothes would be another example – are more 'successful' than things which do not possess such agency and this capacity is inseparable from the ways in which cars are deployed.

The car, then, is a thing of many potentials. It is a social body, an anthropomorphized vehicle, and operates in various modes of spatial negotiation. It can be *wiltja*, windbreak, ceremonial prop, light and music creator, and at the end of its life, may modify the topography where it evokes with sadness for those who knew it in motion, the events, places and people it connected. I have tried to describe something of the richness of the life and death of cars on the Pitjan jatjara Lands. Doubtless there is much more.

With thanks to Gertrude Stotz. Thanks also to Paul Memmott for access to P. Hamilton's paper and fellow students in the Department of Anthropology, University College London for helpful discussion of an earlier version of this chapter.

References

Edwards, B. (1994), 'Mutuka Nyakunytja – Seeing a Motor Car. A Pitjantjatjara Text; Jacky Tjupuru Wankanyja'. *Aboriginal History*, 18:2, Department of Pacific and South East Asian studies, Australian National University, Canberra.

Goddard, C. (1994), *Pitjantjatjara/Yankuntyjatjara to English Dictionary*. Alice Springs: Institute for Aboriginal Development Press,

Hamilton, A. (1979), 'Timeless transformations: Women, Men and History in the Australian Western Desert' (PhD dissertation, University of Sydney).

—— (1987), 'Coming and Going: Aboriginal Mobility in North-West South Australia, 1970–71'. *Records of the South Australia Museum* 20:47–57.

Hamilton, P. (1972), 'Aspects of Interdependence Between Aboriginal Social Behaviour and the Spatial and Physical Environment'. Department of Architecture, University of Sydney.

Healthy Aboriginal Life Team (HALT) (1991), Anangu Way, Alice Springs: Nganampa Health Council.

Kopytoff, I. (1986), 'The Cultural Biography of Things: Commoditization as Process'. In A. Appadurai (ed.) *The Social Life of Things*.

Commodities in Cultural Perspective. Cambridge: Cambridge University Press, pp.64–91.

Lester,Y. (1993), *Yami The Autobiography of Yami Lester*. Alice Springs: Institute for Aboriginal Development AD Press.

Lewis, D. (1976), 'Observations on Route Finding and Spatial Orientation Among the Aboriginal Peoples of the Western Desert Region of Central Australia'. *Oceania,* June 1976 XLVI No. 4, pp.249–282.

Myers, F. (1991a), 'Burning the Truck and Holding the Country: property, time and the negotiation of identity among Pintupi Aborigines'. In *Hunter-gatherers Today*. Ingold, Riches & Woodburn (eds).

—— (1991b), Pintupi Country Pintupi Self Sentiment Place and politics Among Western Desert Aborigines. Smithsonian Institution Press, Washington.

Payne, H. (1984), 'Residency and Ritual Rites'. In J.C. Kassler and J. Stubington (eds), *Problems and Solutions: Occasional Essays in Musicology Presented to Alice M. Moyle*. Place of pubn: Hale and Iremonger, Sydney NSW.

Shivelbusch, W. (1986), The Railway journey: the industrialisation of time and space in the 19th century. Leamington Spa, U.K. New York, NY, USA. Hamburg, Berg.

Spencer, B. & Gillen, F.J. (1899), *The Native Tribes of Central Australia,* London, Macmillan & Co. Ltd.

Stotz, G. (1993), '*Kurdungurlu* Got to drive Toyota: Differential Colonizing Processes among the Warlpiri' (PhD thesis, Deakin University, Australia).

Tindale, N. (1972), 'The Pitjandjara'. In M. G. Bicchieri (eds), *Hunters and Gatherers Today.* New York: Holt, Reinhart and Winston. pp. 217–68.

Wassman, J. (1994), 'The Yupno as Post Newtonian Scientists: The Question of What is "Natural" in Spatial description'. *Man* 29, pp. 645–66.

Notes

1. Anangu means 'people' in the related Pitjantjatjara and Yankuntyjatjara languages but has become a proper noun meaning 'Aboriginal people'. Pitjantjatjara and Yankuntyjatjara people use it to refer to themselves and I

use it in this sense here. I lived mostly on homelands about 35km from Ernabella from March 1997 to August 1998. My fieldwork was enabled by a grant from the Economic and Social Research Council, UK. I gathered information about cars as a part of this research on colour and the senses.

2. 'Tjukur(pa)' is translated as 'Dreaming'.

3. *Inma* means singing and is applied by Anangu both to traditional Aboriginal religion and to Christianity.

4. Returning to visit friends in Ernabella in 1999 I faced difficulty in obtaining a car at the height of the tourist season in Alice Springs. I rented the only hire car I could find which to my horror was literally brand new. No one was impressed by this car (despite its speed and smooth ride) because of its newness and also because it had none of the essential bush features, lacking bullbars at the front, tow bar at the rear and a roof rack, being a four-wheel-drive vehicle built for the metropolitan suburbs.

5. 'Looking after' cars is stated as keeping the oil and water topped up and the battery charged.

6. There are many different kinds of *wiltja* still built but these 'humpies' as Anangu also call them are the commonest now. Peter Hamilton made a comprehensive study of *wiltja* types in Mimili – then Everard Park in 1971–2. P. Hamilton (ibid: 14) points rightly to the visual privacy achieved by the *wiltja* interior with its high contrast between outside and in. Cars give less of such privacy but some new cars with tinted solar glass are appearing on the Lands owned by economically well-off and usually therefore powerful men.

7. Sorry business begins at a death and continues through the funeral, sometimes for months, until a further ceremony, the Opening, takes place. A sorry camp is set up and lived in by close kin of the deceased for the duration of this period away from the place where the deceased habitually lived.

8. Caravans are sometimes used as supplementary housing on homelands but are not towed around as mobile homes.

9. Ngaayyatatjara Pitjanjatjara Yankunyjatjara radio.

10. In September 1999 the centre of Ernabella was laid with bitumen roads in an effort to minimize dust. The centre of Amata, west of Ernabella has had bitumen roads for some time, but generally all roads on the Pitjantjatjara Lands are dirt.

11. Tracking is an essential part of hunting and gathering skills and was one of the few things for which Australian Aboriginal people were admired by the colonists, who thus often employed them as police trackers. Spencer and Gillen (1899: 24–5, *The Native Tribes of Central Australia.* Macmillan and Co Ltd) for example have this to say: 'The tracking powers of the native are well known . . . not only does (he) know the track of every bird and beast, but . . . from the direction in which the last track runs, will tell you whether the animal

is at home or not', and 'the native will recognise the footprint of every individual of his aquaintance'. Today, the tracking ability of Anangu extends to cars and distinguishing the patterns on an individuals shoe sole. Aeroplane tracks across the sky are also noted. The relationship of mark-making to motion then, continues to be important.

12. Some men-only or women-only sites are subject to strict prohibitions and must not be concerned with physically appraoached or, if in open country, looked at by those of the opposite gender.

13. The Stuart Highway is the place where the drivers of bush cars are often stopped by the police for breaking the law for drunk driving and for lacking various statutory car parts, full sets of lights for example. Young men in particular are often in and out of prison for these offences, a situation presently compounded by the Northern Territories' Mandatory Sentencing laws.

14. Stotz (1993: 173–4) has suggested that the combustion engine, particuarly the carburetor, is linked to men's control of fire by the Warlpiri, thus gendering cars as male things.

15. Drawings from the C.P. Mountford Collection State Library of South Australia, collections in the National Museum of Australia, the Ara Irititja Archival Project and Ernabella Arts archive.

The Invisible Car: The Cultural Purification of Road Rage

Mike Michael

Introduction

This chapter concerns a putatively recent phenomenon that has an unexpectedly peculiar relation to the car. The phenomenon is 'road rage', a condition that was thoroughly discoursed in the UK in the early to mid-1990s. Though there are numerous and contested attempts at defining road rage, we can, initially at least, draw upon the UK's Royal Automobile Club's minimal definition presented in their discussion document 'Road Rage' (undated). For the RAC, road rage is 'simply a term to describe a range of anti-social, ill tempered, foolish or violent behaviours by a minority of drivers'. Under the rubric 'anti-social, ill tempered, foolish or violent behaviours' come headlight-flashing, tailgating, cutting in, obscene gestures, obstruction, verbal abuse, running over offending drivers or pedestrians, using various objects to smash windscreens, stabbing with screw drivers and knives, spraying with ammonia, threatening with guns, poking, punching, throttling, beating. The RAC goes on to characterize road rage as entailing the 'altering of an individual's personality whilst driving by a process of dehumanisation' and as 'a total loss of self control' (p2). Accordingly, 'Given the wrong conditions almost anyone can lose their control' (p2). As will be noted below, this is, inevitably, a definition that is 'interested' insofar as it evokes the concerns and strategies of the RAC. Nevertheless, it does highlight the motif of 'loss of control', and links this to the process of driving, indeed, to 'almost everyone's' driving. On the surface, then, it would appear that the link between the car and road rage is pretty straightforward – things happen when driving which 'trigger' road rage.

However, driving is anything but a straightforward process, though it is certainly managed in smooth, seemingly straightforward ways (cf Lynch 1993; Mennell 1995), and the car is anything but a straight-forward manifestation of material culture (cf. Michael 1998, 2000; Lupton 1999; Urry 2000). One of the things that is interesting about road rage is that, as an example of a 'moral panic' (Thompson, 1998), it serves as a condensation of a series of issues and anxieties in contemp-orary Britain. For example, in the print media road rage has been used to express worries about: the state of traffic on the roads and of the transport system in general; recent transformations in British society (people have become more selfish); and the increased levels of all forms of anger and rage (see below) which, it is claimed, are partly prompted by the rise of psychotherapy, with its injunctions to 'express oneself'. But, notice that in all these accounts, the car features barely at all. The car is, it would seem, a rather anonymous part of the immediate and extended setting wherein road rage is enacted.

This 'invisibilization' of the car in the context of media discourses on road rage is what is of interest in the present chapter. In what follows, I will describe instances of how the phenomenon of road rage is, to use Bruno Latour's (1993a) phrase, 'purified'. As such, the human and non-human components that make up road rage are separated – or rather, as we shall see, the role of the human (that is, human nature or human culture) is privileged even while the interconnections between the human and the non-human are tacitly assumed. After briefly reviewing previous work on some of the ways in which road rage has been constructed – that is, purified (especially by the major driving organiziations in the UK) – I will consider a number of discourses that have appeared in the media and that have attempted to contextualize road rage historically, cross-culturally or micro-socially. I aim to show how these contextualizations, while presupposing the hybridity of humans and technologies, simultaneously function to purify such hybridities: that is, the role of the car is downplayed; the car is, as it were, 'invisibilized'. However, I will also suggest that there are places where people do grasp and overtly articulate, albeit in partial and fragmentary ways, the role of the car in road rage. That is to say, there are discourses that minimally formulate the hybridity of the enraged driver. In conclusion, I very briefly explore some of the broader implications of these discourses of hybridity.

Some Initial Observations

John Urry (2000) has recently pointed out that the car has been strangely absent from mainstream sociology and social theory. This neglect is strange because the car has played such a major role in shaping the modern world. Needless to say, there have been various treatments of the car as an exemplar of material culture. Thus it has been seen to be instrumental in the signification of disparate forms of identity – subcultural (Marsh and Collett 1986; Rosengren 1994; Lamvik 1996), gender (Bayley 1986; Hubak 1996; O'Connell 1998) and national (eg Hagman 1994). Moreover, attention has also been drawn to its part in the material-semiotic restructuring of the Western world from the instigation of speed cameras (Stenhoien, 1994) and traffic-calming measures such as sleeping policemen (Latour, 1992) to such now commonplace architectural features as the garage and the motorway (Marsh and Collett 1986; Flink 1988). But there has been no sustained attempt to integrate the car into general sociological accounts of (post)-modernity – at least not until Urry's (2000) notion of 'automobility'.

Part of the reason for this general neglect has been the emphasis upon the 'social' and 'society' in social science (cf Urry, 2000). As Latour (1993a) argues, this division into the human and the non-human, the social and the natural has a long history which he calls the 'Modern Constitution' according to which we moderns have been predisposed to see only these dichotomies. As an example we can point to Bijker's (1995) very useful survey of approaches within the sociology of technology. According to Bijker: '. . . a general pattern can be recognized in which the study of technology and society has been developing. This pattern can, very schematically, be characterized as a sort of slow pendulum movement – a dampened oscillation' (p. 254). First there was outright technological determinism of society, then there was social shaping of technology, then back to a slightly less virulent form of technological determinism until now, barely oscillating in between these extremes, there are a number of approaches that examine the reciprocal determinations of the technological and the social. A particular arena in which one can begin to detect the crumbling of the modern constitution is that of ecopolitics, or what Latour has called political ecology (1997, 1998), where the complex mutualities of the human and non-human, the cultural and the natural – that is, of hybridities – are increasingly actively and self-consciously deployed in the process of doing such politics (cf. Whatmore 1997; Cussins 1997). Further, hints of an occasional grasp and valuation of such hybridities, as we shall glimpse, pervade popular culture (also see Michael, 2000). Such a process of judgement of hybrids is important because, according to Latour, our modernist blindness to them has enabled their sometimes dangerous proliferation, most dramatically exemplified in the increasing pervasiveness of biotechnologies and their products.

However, the key purpose of this chapter is to look at how, in the case of road rage, such mutualities, reciprocalities, hybridities are (un)systematically expunged. Now, there are many discourses that have attempted to characterize, account for, or otherwise explain road rage. Prominent among the actors who provide such discourses are the driving organizations, the Royal Automobile Club (RAC) and the Automobile Association (AA). As longstanding advisors to government (cf. O'Connell 1998) and as providers of much newspaper copy, they are particularly important in shaping popular thinking on road rage. Elsewhere, I have analysed these organizations' contrasting analyses of road rage (Michael 1998, 2000). Suffice it to say in the present context that the RAC takes a largely humanistic or psychotherapeutic stance on road rage: people are dehumanized by virtue of everyday road-use frustrations and an artificial sense of insulation and empowerment provided by the car. As such the typical road-rager would benefit from therapy from a specialist – a process of rehumanization through the recounting of experience and expression of emotion. By comparison, the AA's stance is more behaviourist: road rage is the displacement of stresses and strains through the exercise of animalistic territoriality. The car is an extension of personal territory, and if this is impinged upon aggressive territoriality is triggered. Thus, to avoid road rage one needs to follow a series of behavioural injunctions to diffuse the sense of territoriality: avoid eye contact, do not manoeuvre suddenly, do not rise to the bait, and so on and so forth. What we have, in summary, are two opposing dehumanizations: for the RAC the tendency of the driver is to become more god-like, hubristic; for the AA, it is to become more rat-like, animalistic. Both organizations are engaged in the process of rehumanising – of changing the person, of indeed reinforcing offenders' 'normal' human agency which has been so disastrously transformed within the car.

Note that the car in both cases plays a part – it insulates and empowers, or it is an extension of one's usual personal body space or territory: hybridity is being tacitly assumed. However, in both cases there is also purification: the car and person are extricated from one another. Thus what is primarily being addressed is a 'generic car', or even more basically, a structured space that can be 'empowering' (that is, become part of one's personal powers) or 'territorialized' (in the sense of becoming part of one's personal territory). Indeed, the car is, as noted above, a rather anonymous technological artefact whose rhetorical function in these accounts is to provide a setting in which certain psychological or evolutionary predispositions can be played out. In these accounts a 2CV affords as much empowerment or enables the exercise of as much territoriality as a Mercedes S Class. Despite appearances, then, the car is actually marginal in these accounts – and this is most obvious insofar as it is the human that is the primary point of intervention: Latour's modern constitution finds expression once again.

Now, this differentiation between the RAC's and the AA's accounts are rather fine-grained when compared to the way that road rage is explained more generally in the media. The accounts one finds in many of the extended news-media pieces tend to be 'perambulatory' in the sense of moving through the variety of understandings of road rage with a view to surveying them. However, this is so unsystematic, the accounts are put together, that is, juxtaposed, with such little sensitivity to the formal or intellectual tensions or contradictions between them, that preferable to any notion of 'discursive perambulation' would be 'discursive morassification'. For example, in Esther Oxford's article entitled 'Road Rage' in *The Independent* (25 January 1996) there are presented the following explanatory fragments: overcrowding on the road ('Given that 25 million vehicles clog Britain's roads, it is no surprise that road rage, an aggressive exchange between drivers, is reaching epidemic proportions'); discourteous driving ('Many were annoyed by drivers who cruised in the middle lane, overtook on the inside or were speeding in urban areas'); territoriality; the car as a protected environment; the threateningly animalistic qualities of the car itself ('Dr Marsh regards the design of headlights as particularly problematic. "Headlights give cars an animalistic quality – as do the four wheels. They are rather like horses. Lots of road rage incidents happen because people are not dipping their headlights. The headlights are seen as staring, threatening eyes"'); the primitiveness of our bodily capacities ('Our minds may be sophisticated, but our nervous systems are not'); different personality types (there are two types of people who are likely to succumb to road rage: 'psychopaths' (people with anti-social personality disorders) or 'narcissists' (people who believe that they are special, and that only they know what is 'right')).

What we see here is a process of trawling – a juxtaposition of different explanations drawn from different experts and commentators with no attempt to sort these out, to rank them, to bring consistency to them.

There is, as noted above, a sort of morassification of discourse. But what does this morassification perform? Let me suggest one reading of this unsystematized patterning of explanations. What is performed is the evocation of a choice of causalities, or enabling conditions, of road rage. That is, even if certain causal routes seem inappropriate or incredible for some drivers (e.g. the threatening quality of headlights), there are others which can slot into place to serve as triggers for road rage. This tacit representation of a free market of causalities, a multiplicity of potential causations and possible predispositions, seems to have the rhetorical function of signifying that we are all, *one way or another*, liable to road rage. It enacts the generality of road rage – its pandemic qualities – even if in the specifics of its causalities there is variation. We are all liable to road rage, even if the causal route by which we come to it is personalized.

The foregoing account has actually touched upon the contributory role of the car itself. Not only does it insulate and cocoon, it also has animalistic eyes that challenge and threaten. Beyond this, it is also a piece of technology which, for all its ergonomic sophistication, can become too complicated, too fast, for our nervous system which evolutionarily lags behind. However, while we can detect in this story a presupposition of hybridity, what ends up being emphasized is the problematic limits of people. As with the RAC and AA accounts, what is privileged is the human, indeed, the essential evolutionarily determined, psychologically bounded human. It is this that must be worked around. It is within these evolutionary limits that human agency and self-control must be enabled and exercised. In sum, while there is a tacit assumption of hybridity, it is not in itself addressed – it is curtailed, that is, mediated through commonplace discourses of essential human frailties.

Now, in contrast, there are also accounts which focus not on the changing of people, but on the changing of the technology. There have certainly been programmes funded by the European Union and other agencies assessing the relative merits of such innovations as in-car speed limiters and fully automated speed-control systems as against physical traffic-calming measures and in-car advice systems (for example, Várhelyi, A., Comte, S. & Mäkinen, T. Evaluation of In-Car Speed Limiters, Final Report. MASTER Deliverable 11 (report 3.2.3). Sent for approval to DG VII in September 1998 – http//www3.vti.fi/yki/yki6/master/deliver.htm). Less technologically advanced, more ironic, and perhaps more effective is John Adams' (1995) suggestion that if designers and manufacturers rendered the car less safe, less big, less responsive, less fast, less accurate there would be dramatic decreases in the levels of road rage specifically, and of dangerous driving generally. Indeed, Adams takes this point to an extreme when he remarks 'if all vehicles were to be fitted with long sharp spikes emerging from the centre of their steering wheels (or, if you prefer, high explosives set to detonate on impact), the disparities of vulnerability and lethality between cyclists and lorry drivers would be greatly reduced' (p. 155). Now, these two variants of the technological fix simply reverse the locus of intervention from human to technology. Again, hybridity is assumed and again it is a limited version: technology acts upon a human essence in the sense that, for Adams' solution, less safe cars trigger self-preservational reflexes. In Latour's terms, the human and the non-human are kept separate insofar as there is no 'exchange of properties', as he would put it, in which both person and car come together to produce a different sort of entity that might suggest different sorts of heterogeneous solutions to the problem of road rage (see below).

In the foregoing we have briefly seen how a limited version of hybridity tacitly informs a variety of accounts. It grounds these accounts, but is also purified – articulated through a variety of discourses which reassert modernist dichotomies of human/non-human. One might put it this way: this presuppositional hybridity is 'penumbral': it is neither wholly in the shadow nor in the dark. The significance and function of such penumbral hybridity will be further explored in the next sections where I consider accounts of road rage which attempt to 'contextualize' it. By 'contextualize' I mean to denote a process of comparison with 'similar' behaviours which are situated 'elsewhere'. This 'elsewhere' takes, as far as I can tell, three forms: the elsewhere of history – the past in which earlier versions of road rage took place; the elsewhere of culture – those different societies in which activities comparable to road rage take place; and the elsewhere of what we might call 'microsociety' – those other local social situations wherein are found 'enraged' behaviours that resemble road rage.

To Contextualize is to Universalize

In this section, then, I consider three 'elsewheres' where rage, in various forms, is seen to be enacted. What we shall see is that the contextualization in relation to these elsewheres has the effect of universalizing road rage in the sense of invisibilizing the car – of showing that it is human nature or human culture that lies at the root of such behaviours. In contrast to these rhetorics, I will try to show how technology – in somewhat reformulated form (the car as a heterogeneous, socio-technical network) – is crucial to such displays of road rage.

The Historical Comparison

On occasion we find that road rage is detected in previous epochs – the novelty of road rage is thus denied. I present two examples of such a process of contextualization that serves in the dehistoricizing of road rage. First, a letter to the *Independent* (26 January 1996), entitled 'Oedipal Rage', suggests that road rage was present among the ancients. (That it was thought worth publishing suggests that it was in some way of import)

> From Mr George MacDonald Ross
> Sir: The earliest recorded example of road rage ('Half all drivers are targets of road rage', 24 January) was surely when Oedipus killed his father in an argument over who had priority to drive his chariot over a narrow bridge. Perhaps ancient Greek priority signs were as confusing as ours!
> Yours faithfully,
> George MacDonald Ross
> Leeds
> 24 January

Secondly, Alex Spillius in the Guardian (28 October 1995) draws attention to a reported instance of road rage in the early part of the nineteenth century:

> Indeed the Oldie magazine recently printed an item of 'carriage rage' from 1817. 'Last week I had a row on the road ... with a fellow in a carriage who was impudent to my horse. I gave him a swinging box on the ear, which sent him to the police, who dismissed his complaint,' wrote Lord Byron to Thomas Moore. (Being 'impudent to a horse' was not, by the way, a 19th-century euphemism equivalent to 'sheep scaring'; the other driver merely shouted at Byron's palfrey.)

What both these reflections do is, of course, to downplay the role of the car in the scene of modern day road rage. No longer is it a necessary component in the triggering of road rage – rather, it is the primarily social processes of disputation or impudence that are responsible for these historical instances of road rage. There is, it is implied, nothing historically unique about road rage – it has been a part of travelling since time immemorial. The comparison between these instances and latter-day road rage is accomplished by assuming that what is 'fundamental' – that is, what serves as the criterion which draws these incidents into the same category – about road rage is the eruption of anger and the manifestation of violence by one person upon another. This eruption is triggered by social antagonism. It is not, in this sort of comparative narrative, admissible that these angers and violences might be qualitatively different. That is to say, the car-of-today, with all its complex cultural and material qualities, partially renders the doing of anger and violence historically distinctive. The point is to recover the car-of-today as a heterogeneous socio-technical network that frames a different set of conventions about how one should perceive the acts of others, about what counts as an affront, and about how one should display umbrage (see below).

The Cultural Comparison

In numerous articles we find prefatory remarks on the history of road rage as a phenomenon, or on the genesis of the term 'road rage', in which both phenomenon and term are traced to the United States. Usually, this is done disparagingly. Thus, it is presented as fitting that the phenomenon should first arise in the US – after all, it is in the US where, stereotypically, we (that is the British) would expect, or at least be unsurprised, to find yet another form of chronic violence. Further, that the source of the term 'road rage' is the US is also presented as something to which we (the British) should be resigned: the tacit story is that 'only in America' are such new conditions, syndromes, pathologies so readily identified and named. But there is another dimension to these accounts: once again a product of the US finds its way to the UK. We (the British) are, it would seem, the willing recipients of even the most excessive, the least digestible exports from the US – and this time it is road rage as both behaviour and category. Following Condor (1996) among others, we might suggest that this self-derogation might be linked to the performance of Englishness. The point here is that these sorts of account are about locating road rage in the peculiarity of a particular culture (that of the US) and bemoaning the fact that we are becoming like that culture, even a part of it. The act of 'bemoaning' is the performance of Englishness.

However, I want to examine a different sort of cultural comparison. Occasionally one comes across surveys of the way that 'road rage' is manifested in different parts of the world. In other words, there are accounts of what would trigger road rage in different societies. The article I focus on here is Christopher Middleton's 'It's all the rage of the road' (*Sunday Times*, 25 June 1995), which provides an account of European cultural differences in the triggering of road rage and offers advice to holiday drivers on how to avoid encounters which may lead to the local version of 'road rage'.

Interestingly, before the author looks at the cross-cultural differences in road rage, there is a brief account of road rage in Britain followed by a quote from 'RAC psychologist Conrad King' expounding the 'dehumanization thesis', and according to whom '(dehumanization) is caused by road-use frustrations and an artificial sense of insulation, protection and empowerment provided by the car'. With this in place, the article goes on to provide a 'guide to European road rage' for holiday drivers in which, in order, we have profiles of 'road rage' in Italy, Germany, Spain, France and Greece.

In Italy:

> The one thing that winds an Italian up more quickly than hairpin bends is indecisiveness. A little stutter here, a moment of hesitation there, and the full ensemble of 100 car horns will be the result. 'You have to make up your mind and go,' says the writer Diane Seed, who has lived in Rome for 27 years. 'The important thing is not to catch the eye of another driver; it will be taken as a sign of weakness . . . Italian road rage tends to be noisy, rather than the sneaky French variety.'

In Germany:

> The worst thing you can do to a German is to block his all-conquering path down the autobahn. Because there is no speed limit on the motorways, the only cars that can survive in the fast lane are those that belong to the master race of motor vehicles . . . Middle-lane dawdling is virtually unknown, rebuked not by hooting or shouting but by the ferocious glare of headlights in one's rear-view mirror . . . So disciplined are German drivers that making abusive signs is a criminal offence; it is also unwise, since there are Germans who, in a state of road rage, can lose that methodical reserve. People who have tapped their foreheads to question another driver's mental stability have pulled up at the next service station to find themselves being challenged to a duel.

In Spain:

> If you are going to overtake a Spaniard, first check who it is. If it is a woman, you should be all right. If it is a man on his own, less good, and

if it is a man with his family, forget it. From the average hombre's point of view, being passed by another car is a humiliation, all the more keenly felt if one's wife and children are watching. And the only way to wash off the stain of shame is to take revenge. Usually this will take the form of him re-overtaking you (either side will do) and then slamming on the brakes, secure in the knowledge that if you hit him it will be your fault on the grounds that (a) you were too close and (b) you're a foreigner.

In France:

There is a particular bitchiness to road rage in France which, mixed with four-star anglophobia, forms an explosive combination . . . Paris is the worst place to be seen with British number plates. Whereas indigenous drivers cut up and are cut up with the insouciance of Napoleonic cavalry, it is seen as a slight beyond endurance to have the same done to you by a rosbif in a Rover. Once provoked, the French can be persistent in their desire for vengeance. One English expatriate was recently followed back to her home by an enraged French driver, who camped outside her house shouting anti-British slogans.

In Greece:

The secret of happy Hellenic motoring is to stay constantly on the alert. 'Here everybody is violating every imaginable rule', says the writer Dimitri Mitropoulos . . . Even in the most tense situations, though, Greek road rage rarely boils over into violence. 'Because we let off steam on a regular basis, there is not the enthusiasm for fighting', says Mitropoulos. 'You may often see two men arguing, but they won't be hitting each other.'

I have presented these comparisons in some little detail because, in an albeit condensed way, they do convey some of the different – stereotyped – behaviours which are being brought under the rubric of 'road rage'. The key issue is what makes these stereotypes comparable – the same but different. Minimally, these rages all take place in relation to roads and cars, but the triggering mechanisms are somewhat different. Or rather, in each country different cultural values pertain: the mores, realized in the social conventions of what is to count as an affront and what a commensurate riposte, vary. Thus, in Italy it is hesitancy that triggers noisy horn-blowing protest, in Germany it is slowness that triggers headlight flashing and tailgating, in Spain it is overtaking that prompts re-overtaking and sudden braking, in France it is being English and behaving like the French that necessitates vengeance, in Greece it is inappropriately following the rules that demands histrionic argumentation.

What the article accomplishes, then, is the assertion of cultural difference. The car is again invisibilized: it is effectively simply another setting in which 'national character' is made manifest. What is missed out of this account is the fact that the car as complex part of a complex heterogeneous socio-technical network (e.g. Latour 1987) that incorporates, and is incorporated by, 'national character'. To put it less ironically, the cultural conventions which structure the production of 'road rage-like' behaviours are related to the car itself, where the car is conceptualized as a distributed material-semiotic 'nexus' (Whitehead, 1929) or hybrid collectif (Callon and Law, 1995) that mediates, and is mediated by, local and global cultural and material conditions. The behaviours that are associated with the car are thus locally distinctive, but that localness has in part been mediated through the car which comes to have a local history, but also affects that local history. Moreover, that local or national history vis-à-vis the car is being worked out against other national histories. This, of course, is exactly what is being performed in the Middleton article quoted above. Thus, the car is embroiled in the (re)production of local cultural conditions, in part through processes of cross-cultural comparison.

But further, we should also recall that the car is an entity that, as both material and sign, flows cross-nationally and cross-culturally (Appadurai, 1990; Lash and Urry, 1994; Urry, 2000). As a material-semiotic artefact it is has been a means through which, according to Sachs (1984), the distinctly modernist predilection for speed and competition has been, in part, disseminated, globalized (see also Virilio 1977, 1995). It is these modernist predilections – heterogeneously embodied in the car as a series of conventions or scripts (see Akrich 1992) – that resource road rage, not some ahistorical, over-psychologized predisposition of feeling empowered in 'generic' metal boxes on wheels. Thus, the car itself has played a part in the globaliziation of 'road rage type behaviours' but this is in its complex, historical, *conjoint* capacity as a signifier of speed (and it has from outset been advertised in the terms of competition – from its beginnings car racing was seen as means of promoting the car – cf. O'Connell 1998) and material entity with certain physical capabilities (greater power, more responsivity, better road-holding, etc.). To reiterate, these globalized potentials are, however, realized in – articulated with – the local.

The Micro-social 'Behavioural' Comparison

Another set of comparisons again contextualize road rage, this time in relation to different forms of rage. Thus, we find rage not only among drivers, but also among cinema-goers, pedestrians, golfers. A particularly good example of this comparison across micro-social settings is provided in Roger Tedre's article 'Aggression: Suddenly, spectacularly, losing your cool is all the rage' in the *Observer* (15 October 1995). The by-line 'From cinemas to golf courses, unprecedented levels of stress in modern life are prompting more confrontations in everyday situations' immediately sets up the fact that it is 'unprecedented levels of stress in modern life' that result in the enraged displays which are to be listed, as if these 'unprecedented levels of stress in modern life' were somehow separate from the technologies that inhabit 'modern life' with us.

After a number of lurid examples of rage in a cinema, in a supermarket queue and in the process of walking along the pavement, Terdre summarizes with a list of different types of rage:

Little things that irritate

Road rage: Aggressive behaviour behind the wheel, headlight-flashing, obscene gestures, and verbal and physical abuse.

Trolley rage: Often sparked by small children, slow unpackers and change-fumblers. Queue jumpers have felt the force of a fist.

Cinema rage: Slow and sotto voce, but erupts spectacularly because of the confined location. Most tall people have been unwittingly on the receiving end.

Pavement rage: A US import, typified by aggressive comments and shoving.

Phone rage: Brought on by answering machines, voice mail and inefficient receptionists.

Golf rage: The latest rage has hit the sport of gentlemen. Caused by slow players who do not allow faster and more experienced ones to go on ahead.

In addition to these there are a variety of other rages available: cycling – rage at pedestrians or drivers who are oblivious to cyclists; car alarm – rage at car alarms that go off unprompted or keep going off at intermittent intervals; tube – fury at the inefficiency of, and chronic tardiness within, the London Underground railway system; computer – anger when the computer breaks down, or is too slow; rod – examples where anglers have turned on another whose dog had unsettled their fishing water; air – where aeroplane passengers become abusive toward airline staff or fellow-travellers.

Now, in these examples comparability is rendered across different 'forms' of rage by virtue of, as the by-line indicates, the 'unprecedented levels of stress in modern life'. These 'levels of stress' serve as the conditions by which we have become primed to 'go off': the slightest affront (or insult or inconvenience) and we become angered. Once more the technology is assumed to play a role but it is that of, it seems, both a partial trigger and partial setting. The hybridity is a limited one. If we were to think through these versions of rage in terms of hetero-geneity, we might note that, on a general level, the commonality across these micro-social settings is one partly grounded in technologies which are a part of a socio-technical network in which time has become accel-erated (Adam 1998; MacNaghten and Urry 1998), and wherein 'speed' is increasingly prominent (Millar and Schwarz, 1998). These techno-logies are, therefore, part of the network conditions of possibility for these 'unprecedented levels of stress in modern life'. This network includes mundane technologies such as cars and paving stones as well as those supposedly 'epochal' nanosecond information technologies. The point is that, in the above newspaper article, the role of these variegated technologies of speed (and the expectation of speed) in the generalized 'unprecedented levels of stress in modern life' has become obscured by the stress upon stress.

We might further propose that these 'technological settings', or these technological artefacts – car, fishing rod, bicycle, cinema, computer, pavement, telephone, shopping trolley – as networks, as nexûs, would entail a 'promise' that is inscribed in them materially and semiotically. At minimum, such promises would entail that the technology should not break down; neither it nor its user should be 'unnecessarily' impeded (cf. Michael, 2000, 1998). If impediments arise, are we right to see the reactions to these as of a piece across these different technological artefacts and their respective socio-technical networks? I would answer 'yes' and 'no': as we have seen, there might be something of a generalized, perhaps 'grammaticized' (cf. Michael 1996; Mulhausler and Harré 1990) modernist 'ethos of speed', but each actualization of this 'ethos' will be mediated by the local, micro-socio-technical conditions. In other words, there will be different qualities as well as quantities of speed attached to these conditions. These different rages in their different socio-technical settings reflect not simply a response to a generic transgression – the very quality of the rage reflects its local setting. This, however, all needs to be empirically investigated.

In this section, I have listed a number of ways in which, in accounting for road rage, the hybridity of cars and persons is at once assumed and filtered through modernist dichotomizing or purifying discourses in which the human emerges as the explanation of road rage. Further, we have seen that this 'human' takes several forms. On the one hand there are specific and general human psychological predispositions (e.g. the feelings of insulation inside a car); on the other, there are cultural conditions whose horizons are extended temporally (across epochs) and spatially (across nation-states and micro-social settings). All I have done is to suggest that technology in general, and the car in particular, have constitutive roles to play too.

However, in my zeal to re-assert the role of technology – to give it equal footing with the cultural and the psychological – in such enraged behaviours I have perhaps swung too far in the opposite direction and underplayed the role of the human. What is needed is a way of articulating how the local and generic human are, like the local and generic (that is, modernist) car, mutualistically crucial in any worthwhile account of road rage. Accordingly, and in line with a number of approaches that speak of monsters (e.g. Law 1991), (low-tech) cyborgs (cf. Gray 1995) and material culture (e.g. Dant 1999), I now consider road rage explicitly in terms of human–non-human 'hybridity'.

The Rage of Hybrids

In the foregoing I have focused especially on the modernist process of purification in order to suggest ways in which hybridity, which we have seen assumed in a number of texts, might be directly addressed. As we might expect, however, there are instances where driving and road rage have been, at least in part, openly articulated in hybridic terms. So, running in parallel with the assumption – and purification – of the hybridity of road rage are commonplace partial articulations of its hybridity. Lupton (1999) detects instances of this when she suggests that people-in-cars tend to treat one another as 'cyborgs'. Thus, in her analysis of a number of drivers' accounts, she detects an 'elision of human/machine' (p. 64). As she puts it in her commentary on one respondent's account: 'The man describes his own car as if it were his own body that is being roughly handled: "He just gave me a shove in the back"' (p. 64). However, this elision might have as much to do with the available metaphors we live by (Lakoff and Johnson, 1980) as with the emergence a *bona fide* hybridic vocabulary: the car has a 'back' too, after all.

More suggestive are the more tortuous and fragmented articulations in which, as I have noted elsewhere (e.g. Michael 1998), there does seem to be a developing language of hybridity. Thus, an Austin 1100 home page editorial (http://www.users.dircon.co.uk/-canstey/1100_ed5.htm – 18 August 1996) identifies a combination of car and person that is especially prone to road rage. This hybrid is made up of the modern car (eg 'Vauxhall Vomitra' (Vectra), 'Ford Mundano' (Mondeo) and 'Renault Mogadon' (Megane)) which is a 'deeply unpleasant place to be' – full of cheap plastic and synthetic seating materials that 'assault every sensibility'. Seated at the wheel of this modern car is a particular human figure, 'Rupert Rep' – an archetypical sales*man* figure – who is measured by his 'Total Plonker index' (where 'plonker' is British vernacular for a foolish, idiotic person). Together, these make up the most dangerous of monsters (Law 1991).

Let me now turn to a 'simple phrase' that one comes across again and again when road ragers attempt to characterize what the doing of road rage entails. A close 'reading' of this 'simple phrase' suggests that hybridity can be detected underpinning the most mundane accounts of road rage. In paraphrase this mini-narrative takes the following general form: 'When I get behind the wheel of a car, I am a completely different person.' A good example of this is taken from the TV documentary 'Inside Story: Road Rage' (BBC1, 28 June 1995): 'Now when I get into a car, my character changes . . . that sometimes I can't believe the things that I say, just come out of my mouth the way they do.' What this implies is a complex layering of discourse in which, in the course of what is effectively a simple excuse, there is an attempt to redistribute agency and reconfigure enabling conditions (cf. Scott and Lyman 1968). So, on the one hand, this statement can be regarded as an excuse – a shifting of responsibility where one is transformed from being a good, rational 'civilized' person into a foul-mouthed, aggressive menace. One's essence has been shown to be fluid. The speaker changes on coming into contact with the car. However, for this excuse to work, the car must remain a constant, an obdurate entity. Moreover, this obduracy is associated with cars-in-general: the use of the indefinite article ('Now when I get into a car . . .') suggests that cars-in-general have certain common properties that predisposes the driver to bad behaviour. But, this excuse can only function against a tacit recognition, or rather a neglect, of the culturally available knowledge that cars are also designed to 'calm' one, as well as to 'enthuse' one. So, on the other hand, cars have an ambiguous identity (or contain ambivalent scripts, to draw again on Akrich's (1992) terminology): as Urry (2000) has remarked the car is both a racing machine (Porsche) and a safe family vehicle (Volvo). Naturally, these dual characteristics pertain not merely to the car as a singularity, but in its emergence within a complex socio-technical network. As culturalized material and materialized culture, the car prescribes *and* proscribes speed, competition (and thus the possibility of the performance of 'loss of control'); semiotically and materially, it enables *and* disables sensible, safe, cautious comportment. Thus, the essence of the car is itself changeable – its apparent obduracy is grounded in a marginalization of fluidity, of ambiguity. Now, it is the human actor who is obdurate, a singularity able to exercise agency in determining which 'version' of the car to actualize (Warner 1986). To put it another way, if on one side of the discursive coin we find 'Now when I get into a car, my character changes . . .', on the other we find 'Now when I get into a car, the car changes . . .'. More formally, drawing on Billig's rhetorical terms, the meaning of the former account only makes sense in the context of the latter. Billig phrases this point in the following way: 'The context of opinion-giving is a context of argumentation. Opinions are offered where there are counter-opinions. The argument "for" a position is always an argument "against" a counter-position. Thus the meaning of an opinion is dependent upon the opinions which it is countering . . .' (Billig 1991, p. 17). In the context of the modernist constitution, the 'opinion' 'Now when I get into a car, my character changes . . .' is countering the opinion 'Now when I get into a car, the car changes . . .'.

As Latour (1993b, 1999) has noted, these opposing accounts find parallel expression in relation to various human-technology inter-actions. In reflecting upon the reasons put forward for various killings in the United States, Latour, identifies two contrary views which neatly parallel the two phrases examined above: 'it is guns that kill' (people are at the effect of guns) and it is 'people that kill' (guns are at the effect of people). Rather than ascribing essences to the 'gun' and the 'citizen' – each being either good, or bad, or neutral, what Latour aims to do is to show how a new hybrid emerges – the citizen-gun – that entails new associations, new goals, new translations and so on. As one enters into an association with a gun, both citizen and gun become different. As Latour (1993b) puts it: 'The dual mistake of the materialists and of the sociologists is to start with essences, either those of subjects or those of objects . . . Either you give too much to the gun or too much to the gun-holder. Neither the subject, nor the object, nor their goals are fixed for ever. We have to shift our attention to this unknown X, this hybrid which can truly be said to act' (p. 6). So, here, what should be judged is not the gun or the person (or in the present case, the car and the driver) – not object or subject alone – but the combination, the hybrid. And of course, as we have noted, neither the citizen nor the gun, driver nor car is itself pure, respectively pure human and pure machine. They come to each other as hybrids – in the concrete circumstance of their meeting, they move through and, are moved by, other humans and non-humans which are likewise hybrid.

We have caught a glimpse of this judgement of hybridity in the Austin 1100 editorial cited above. However, this differs from Latour's account: the editorial assumes respective essences for the driver and the car that comprise the road-rager: the car is in essence an unpleasant, aggravating place to be in (hence 'Vomitra' for Vectra) and the driver is a sales representative with a 'total plonker index'. Nevertheless, there does seem to be emerging a proto-judgement (evidenced not least in the present text) of hybrids – of cars-and-people. But a question that arises in light of this development is: who/what is doing the judging?

Concluding Remark: Judging Hybrids

The writers and speakers who have in the foregoing, in their various ways, described, characterized, accounted for, explained, purified and judged road rage have been, by and large, unproblematic. That is to say, these figures have themselves been purified: in the present account, they are pure humans who pass judgement. However, if we take seriously that humans are always already integrated into socio-technical networks, we should regard these statements not simply as issuing from pure humans, nor even as local instantiations of discourses or ideologies or narratives, that is, of 'the purified cultural'. Rather, we need to see these judgements as the product *of* hybrids, of socio-technical networks, of heterogeneous relations.

In light of this, a general and overly schematic comment is in order. There is a need to see these speaker/writers – these enunciators of judgement – as parts of networks in which, semiotically and materially (that is, impurely), purified humans and machines are on-goingly reproduced. Thus the enunciator is embroiled in networks in which they are constantly being performed, as Callon and Law (1995) would put it, as pure, singularized humans. From the huge array of cultural resources that go into making up the 'sovereign individual' (Abercrombie,

Hill and Turner 1986) to the mundane technologies of cash-dispensing machines, cars and turnstiles, hybrids are being at once constituted and purified, disentangled into humans and machines. When someone judges a hybrid, this is no less the action *of* a hybrid. What is needed, then, is a vocabulary, a range of conceptual tools, that can serve in the re-writing of this chapter in terms of hybrids, where the stories, excuses, accounts and comparisons can be addressed as road-rage discourses performed by hybrids, not least by the hybrids that perform road rage itself.

References

Abercrombie, N., Hill, S. and Turner, B. (1986), *The Sovereign Individual*. London: Allen & Unwin.

Adam, B. (1998). *Timescapes of Modernity*. London: Routledge.

Adams, J. (1995), *Risk*, London: UCC Press.

Akrich, M (1992), 'The de-scription of technical objects', in W.E. Bijker and J. Law (eds), *Shaping Technology/Building Society*. Cambridge, Mass.: MIT Press.

Appadurai, A. (1990), 'Disjuncture and difference in the global gultural economy', *Theory, Culture and Society*, 7, 295–310.

Bayley, S. (1986), *Sex, Drink and Fast Cars*. London: Faber & Faber.

Bijker, W.E. (1995). 'Sociohistorical technology studies'. In S. Jasanoff, G.E. Markle, J.C. Peterson and T. Pinch, (eds), *Handbook of Science and Technology Studies*. Thousand Oaks, Cal.: Sage.

Billig, M. (1991), *Ideology and opinions*. London: Sage.

Callon, M. and Law, J. (1995), 'Agency and the hybrid c*ollectif*'. *The South Atlantic Quarterly*, 94, 481–507.

Condor, S. (1996), 'Unimagined Community? Some Social Psychological Issues Concerning English National Identity'. In G.M. Breakwell and E.L. Speri (eds), *Changing European Identities: Social Psychological Analyses of Social Change*. London: Butterworth-Heinemann, pp. 41–68.

Cussins, C. (1997), 'Elephants, biodiversity and complexity: Amboseli National Park, Kenya'. Paper presented at Actor-Network Theory and After Conference, Keele University.

Dant, T. (1999), *Material Culture in the Modern World*, Buckingham: Open University Press.

Flink, J.J. (1988), *The Automobile Age*. Cambridge, Mass.: MIT Press.

Gray, C.H. (ed.) (1995), *The Cyborg Handbook*. New York: Routledge.

Hagman, O. (1994), 'The Swedishness of cars in Sweden'. In K.H. Sorensen (ed.), *The Car and Its Environments*, The Past, Present and the Future of the Motor Car in Europe. Proceedings from the cost A4 workshop in Trandheim, Norway. Published by the European Commission.

Haraway, D. (1991), *Simians, Cyborgs and Nature*. London: Free Association Books.

Hubak, M. (1996), 'The car as a cultural statement: car advertising as gendered socio-technical scripts'. In M. Lie and K.H. Sorensen (eds) *Making Technology Our Own? Domesticating Technologies into Everyday Life*. Oslo: Scandinavian University Press.

Lakoff, G. and Johnson, M. (1980), *Metaphors We Live By*. Chicago: University of Chicago Press.

Lamvik, G.M. (1996), 'A fairy tale on wheels: the car as a vehicle for meaning within a Norwegian subculture'. K.H. Sorensen and M. Lie (eds), *Making Technology Our Own? Domesticating Technologies into Everyday Life,* Oslo: Scandinavian University Press.

Lash, S. and Urry, J. (1994), *Economies of Signs and Space*. London: Sage.

Latour, B. (1987), *Science in Action: How to Follow Engineers in Society*. Milton Keynes: Open University Press.

—— (1992), 'Where are the missing masses? A sociology of a few mundane artifacts'. In W.E. Bijker and J. Law (eds), *Shaping Technology/ Building Society*. Cambridge, Mass.: MIT Press.

—— (1993a), *We Have Never been Modern*, Hemel Hempstead: Harvester Wheatsheaf.

—— (1993b), 'On Technical Mediation: The Messenger Lectures on the *Evolution of Civilization*'. Cornell University, Institute of Economic Research: Working Papers Series.

—— (1997), 'A few steps toward an anthropology of the iconoclastic gesture'. *Science in Context*, 10, 63–83.

—— (1998), 'To modernise or ecologise? That is the question'. In B. Braun and N. Castree (eds), *Remaking Reality: Nature at the Millenium*. London: Routledge.

—— (1999), *Pandora's Hope: Essays on the Reality of Science Studies*. Cambridge, Mass.: Harvard University Press.

Law, J. (1991), 'Introduction: monsters, machines and sociotechnical relations'. In Law, J. (ed.), *A Sociology of Monsters*. London: Routledge.

Lupton, D. (1999), 'Monsters in Metal Cocoons: "Road Rage" and Cyborg Bodies'. *Body and Society* 5(1): 57–73.

Lynch, M. (1993), *Scientific Practice and Ordinary Action: Ethnomethodology and Social Studies of Science*. Cambridge: Cambridge University Press.

MacNaghten, P. and Urry, J. (1998), *Contested Nature*. London: Sage.

Marsh, P. and Collett, P. (1986), *Driving Passion: The Psychology of the Car*. London: Jonathan Cape.

Mennell, S. (1995), 'Comment on technicization and civilization'. *Theory, Culture and Society*, 12, 1–5.

Michael, M. (1996), *Constructing Identities: The Social, the Nonhuman and Change*. London: Sage.

—— (1998), 'Co(a)gency and Cars: The Case of Road Rage'. In B. Brenna, J. Law and I. Moser (eds), *Machines, Agency and Desire*. Oslo: TVM, pp. 125–41.

—— (2000), *Reconnecting Culture, Technology and Nature: From Society to Heterogeneity*. London: Routledge.

Millar, J. and Schwarz, M. (1998), 'Introduction – speed is a vehicle'. In J. Millar and M. Schwarz (eds), *Speed – Visions of an Accelerated Age*. London: The Photographers' Gallery and the Trustees of the Whitechapel Art Gallery.

Mulhausler, P. and Harré, R. (1990), *Pronouns and People: The Linguistic Construction of Social and Personal Identity*. Oxford: Blackwell.

O'Connell, S. (1998), *The Car in British Society: Class, Gender and Motoring 1896–1939*. Manchester: Manchester University Press.

Rosengren, A. (1994), 'Some notes on the male motoring world in a Swedish community'. In K.H. Sorensen (ed.), *The Car and Its Environments*. Brussels: The Past, Present and the Future of the Motor Car in Europe. Proceedings from the cost A4 workshop in Trandheim, Norway. Published by the European Commission.

Sachs, W. (1984), *For Love of the Automobile*. Berkley: University of California Press.

Scott, M.B. and Lyman, S.M. (1968), 'Accounts'. *Americal Sociological Review*, 22, 46–62.

Stenoien, J.M. (1994), 'Controlling the car: a regime change in the political understanding of traffic risk in Norway'. In K.H. Sorensen, (ed.) *The Car and Its Environments*. Brussels: The Past, Present and the Future of the Motor Car in Europe. Proceedings from the cost A4 workshop in Trandheim, Norway. Published by the European Commission.

Thompson, K. (1998), *Moral Panics*. London: Routledge.

Urry, J. (2000), *Sociology Beyond Societies*. London: Routledge.

Virilio, P. (1977/1986), *Speed and Politics*. New York: Semiotext(e).

—— (1995), *The Art of the Motor*. Minneapolis: University of Minnesota Press.

Warner, C.T. (1986), 'Anger and similar delusions'. In R. Harre (ed.), *The Social Construction of Emotions*. Oxford: Blackwell.

Werbner, P. and Modood, T. (eds) (1997), *Debating Cultural Hybridity*. London: Zed Books.

Whatmore, S. (1997), 'Dissecting the autonomous self: hybrid cartographies for a relational ethics'. *Environment and Planning D: Society and Space*, 15, 37–53.

Whitehead, A.N. (1929), *Process and Reality*. Cambridge: Cambridge University Press.

Driving While Black

Paul Gilroy

'One of the first things I did with my profits was to buy myself a car, a '77 Cutlass Supreme four-door I got off one of the homeboys from Compton for three hundred dollars. It didn't drive for shit, and every time you hit the brakes you could hear it squealing like a bitch in heat, but I loved the lines of that car and the way I looked sitting behind the wheel . . .'

Snoop Dogg

Looked at from right outside, the traffic flows and their regulation are clearly a social order of a determined kind, yet what is experienced inside them – in the conditioned atmosphere and internal music of this windowed shell – is movement, choice of direction, the pursuit of self-determined private purposes. All the other shells are moving, in comparable ways but for their own different private ends. They are not so much other people, in any full sense, but other units that signal and are signalled to, so that private mobilities can proceed safely and relatively unhindered. And if all this is seen from the outside as in deep ways determined, or in some sweeping glance as dehumanised, that is not at all how it feels inside the shell, with people you want to be with going where you want to go.

Raymond Williams

The twentieth century was the century of the automobile, of auto-mobility and mass motorization. Commerce in motor vehicles still constitutes the overheated core of unchecked and unsustainable consumer capitalism, but the impact of car culture extends far beyond those buoyant commercial processes. Everywhere cities and civic inter-action have been transformed absolutely. Novel and damaging patterns

81

created by motorization have profoundly altered the political economy of everyday life. In an accelerating world, newly illuminated by electricity, cars could appear recurrently as pre-eminent symbols of power and prestige. They were manifestations of wealth as well as elements of wholly unprecedented personal autonomy. Using them established potent criteria for beauty, prestige and style; ritualized entry into adulthood; and affected the psychologies of gender and generation. Automobiles transformed the meaning and the experience of property in relation to both time and value.

It is significant that these machines, which are probably the most destructive and seductive commodities around us, were originally the toys of rich men. The pleasures they offered soon took them beyond their original constituency and, as prices fell sharply, cars and their cultures provided a cue to reorder and reorganize social relationships – even those constituted through and around the deadly divisions of 'race' and its hierarchies.

Given the epoch-making impact of the automobile, there are surprisingly few discussions of it in histories of black Atlantic and African-American vernacular cultures, political mentalities and freedom aspirations. The motor car and the social and cultural relations it created and reinforced are – as in so much twentieth-century social science, even in those specialized studies that purport to address space, public culture and city life – rapidly passed over, naturalized or simply ignored. This chapter can only stir up sediment, but it aims at more than a reflection on those manifold missed opportunities. It ponders the uniquely intense association of cars and freedom in black culture and suggests, first, that the motor car is a bigger issue than historians of the black vernacular and subaltern social life have let it be. Secondly, it argues that taking cars into account in the lives of black communities and analyses of their broadest political and economic hopes has become unavoidable because an assessment of how these groups have responded to and participated in consumer culture is now imperative. Having faced that obligation, it asks whether any new insights might be deduced from better understanding of the special 'symptomatic' points where cars acquired an importance far beyond their material and even their symbolic political currencies.

One interesting clue to the puzzle these symptoms produce lies in the fact that the vernacular speech of African-Americans, still sometimes refers to cars as 'whips'.[1] Deeply repressed and fragmentary acknowledgements of the slave past may be quietly active here, undergirding the patterns of sometimes ostentatious and even excessive market behaviour associated with black consumerism in general and the African-American love of automobiles in particular.[2] The data on African-American consumer behaviour is contradictory and has not as yet been subjected to a detailed historical periodization (Dusenberry 1949; Weems 1994; Molnar and Lamont forthcoming). So I tentatively suggest this possibility, as a bemused non-driver who has been perplexed by the distinctive social relations produced in the USA by extensive corporate and governmental manipulation of the global market in petroleum, by the lobbying power of oil companies and car manufacturers against public transport and by the capacity of the motor manufacturers to address their black consumers through a curious blend of separatist and black nationalist rhetorics with the routine themes of American automotive utopianism (Nadis and MacKenzie 1993; Kay 1997).[3]

To pursue a critical angle on African-American car cultures does not mean that I play down the relative poverty of African-Americans and the entrenchment of their continuing segregation, or ignore the sharp class divisions inside their communities, or minimize the authoritarian force of that stern American obligation to drive and lead a driving life. Those important factors make the enigma of African-American auto-consumerism all the more profound. It is therefore important to underline that this eco-political speculation over aspects of African-American consumption does not pretend or aspire to be an analysis of African-American consumer behaviour as a whole. My concerns are of a different, more restricted order and I can clarify them immediately through the admission that they have been provoked not only by my distaste at the apparent triumph of anti-political and assertively immoral consumerism in black popular culture but also by difficult insights of Adorno into the deep cultural processes whereby 'the triumph of mounting mileage ritually appeases the fear of the fugitive' (1974: 102).

For all we know, Adorno may have been nostalgic for earlier phases of 'automobilism' when motor vehicles were required to follow a man with a red flag and were not allowed to exceed four miles an hour on public highways. Watching the USA through the lens of his exile's life in California, he saw the normative place of walking start to fade away and the liberal era that could be defined by the rhythm of the bourgeois promenade displaced by the altogether different tempo supplied largely by wholesale resort to the private car. Struggling with these difficult problems in the period before joggers became a familiar sight in cities and their suburbs, Adorno noted some of the ways in which the disturbance created by running in the street might bear witness not only to the past, primal terrors of our species but to the continuing vulnerability of frail humanity in harsher urban environments that have become inhospitable to the extent that walking has died out there. Any residual terror could be effectively mastered by 'deflecting it from one's own body and at the same time effortlessly surpassing it' in what were then the emergent cults of speed and sport. This in turn seems to suggest that the destructive automobile and its undoubted pleasures can be understood in the historic setting supplied by traditions of flight, restlessness and mobility that characterized American society in its transition to mass automotivity and the new conceptions of speed and space that derived from it.

Adorno's enigmatic observations might also be employed to reflect upon the particular responses of African-Americans whose histories of confinement and coerced labour must have given them additional receptivity to the pleasures of auto-autonomy as a means of escape, transcendence and perhaps even resistance. However, I should repeat that this tentative consideration of cars within the wider framework of America's racial politics is not a means to present US blacks as special dupes of consumerism. Instead, it raises the provocative possibility that their distinctive history of propertylessness and material deprivation has inclined them towards a disproportionate investment in particular forms of property that are publicly visible and the status that corresponds to them.

The automobile appears, then, at the very core of America's complex negotiations with its own absurd racial codings. African-American car cultures can certainly signify that the official scripts of respectable domesticity and deferred gratification have been rejected while, at the same time, they suggest that that dream is already being surpassed, overtaken, by more powerful and reckless desires that can be examined through the predisposition to what the marketing literature calls 'status purchasing'. There can be no doubt that cars, which are less easily counterfeited than similarly branded watches, clothing and luggage are now at the heart of the cultures of compensation with which some American blacks have salved the chronic injuries of racialized hierarchy.

We need to be able to understand the weak and fading patterns of resistance or struggle that are being articulated here. However, it would also be wrong to overlook the possibility that the same flamboyant gestures of disavowal from the informal regulations of the colour-coded mainstream are being combined with other factors. Their fleeting negativity betrays a semi-transgressive desire to join in the carnival of American plenitude as full participants in ways that racism might deny.

Their accommodation with and subordination to power is accomplished by another, related yearning: an unsubversive will to triumph in the game of consumerism and to make consumer citizenship and its brand identities eclipse the merely political forms of belonging promoted by governments and political processes. In securing this triumph for capitalism, the totemic, sublime power of the motor car demands fresh consideration of the ways in which both the resistance and the resignation of black political interests are now to be understood and evaluated.

The determinedly critical standpoint outlined here is offered as a route into a more motivated political history of car cultures than is currently available. It asks, for example, whether a hostile discussion of cars and their cultures might be employed to help clarify and perhaps rethink the very nature of black political formations at this pivotal, post-colonial stage in their planetary development. This is not only because a critical assessment of cars and car culture can yield tough but productive questions about the notions of progress, development, growth, wealth and prosperity and their relation to unchallenged and apparently invulnerable industrial, post-industrial and commercial capitalisms. The central issues come into focus once we remember that though automobiles are global phenomena, access to them is far from universal. Where cars are available only in limited quantities and remain a luxury item, a big, noisy, polluting badge of relative prosperity, they will be insufficient to orchestrate a car culture, or energize the malign anti-sociality of what we are required to call car-based civilization. However, where they come to saturate social and economic relations, deforming the idea of public good in the process, even among minority populations automobility can pose fundamental problems of solidarity and translocal connection. This is an especially acute danger for black communities that cannot be confined locally and who must seek new ways of becoming present to one another amidst the techno-cultural ferment of the information age.

We shall see in a moment that among blacks in many parts of the world, the car and their attitudes toward it provide complicated tests of consumer culture and its power to depoliticize, disorient and mystify. Analysis of car use can illuminate the tacit enforcement of segregated space that is a growing feature of metropolitan life inside and outside the US. Automobility also helps to show up the deepening lines of class division inside racialized communities that are now anything but spontaneously homogenous or unified.[4] Tracking changing *attitudes* toward the car, through good and bad times, recession and expansion, resurgence and defeat can also be instructive. The private car is an index of hegemony. The failures of public transport confirm and cement the appeal of radically individualistic solutions and communicate the general defeat of the idea of public good. Seen in this way, the motor car is far more than a mere product, an inert commodity or a neutral piece of innocent technology. It acquires a special force, and becomes a social and political actor that shapes the industrial and de-industrializing worlds through which it moves even as it damages both them and us.

I want to go further still and argue that the car's power can be used to illuminate some neglected aspects of the twentieth century's black freedom struggles. These were social movements animated by people who had very good reasons to fear the constraints involved in being tied to one place and who were, as we shall see, especially drawn to the allure of speed, autonomy and privatized transport quite apart from their attraction to the automobile as a provocative emblem of wealth and status. Their enthusiasm for the car and the subsequent inability to see beyond its windscreen reveal how those movements and their conceptions of freedom have been transformed, compromised, distracted and diverted. Today they have helped to deliver us to a historic point where blackness can easily become less an index of hurt, resistance or solidarity in the face of persistent and systematic inequality than one more faintly exotic life-style 'option' conferred by the multi-cultural alchemy of heavily branded commodities and the pre-sealed, 'ethnic' identities that apparently match them. From this perspective, freedom often entails little more than winning a long-denied opportunity to shop on the same terms as other, more privileged citizens further up the wobbly ladder of racial hierarchy. In other words, it would appear that a significant measure of respect and recognition capable of mediating or reversing the effects of subordination can now be simply bought or at least simulated. The desired social effects are to be conferred on purchasers by objects that they own, use or display. In their own eyes and perhaps also in the eyes of others, these subjected people become different at this point of branded visibility. Beyond functionality, use value and all metaphysics of utility (Baudrillard 1981 esp. ch. 7), these users park themselves in a field between two poles. The first we can call shopper's rebellion: here the 'official' value given to these prizes by a world of work and wages is supposedly altered or at least ironically commented upon in a counter-axiology which may become quite elaborate. The second position, defined most obviously by the moods of anxious individuals who want to answer the impact of racism on their lives by buying in rather than dropping out, accepts these objects as a means to seem wealthier, prouder and thus more respectable. Both of these responses have produced high-risk strategies where they accept the link between commodities and identities. It is not only that the resulting forms of attention can release violent envy and compound the petty hatreds of a violently segregated world. The historian David Nye has pinpointed some of the tragic consequences that can follow when this intimate, transient bond is carried over into the world of automotivity: 'to the extent that the buyer invested personal meaning in a car, its obsolescence underlined how unstable the sense of identity can be when underwritten by consumption' (1998: 182).

With these possibilities in mind, it should be easier to see that the car and car-worship distill all the moral and political difficulties of consumerism into pure and potent forms. Some of the liberatory aspects of African-American car use have been revealed, for example, in bell hooks' striking memoir of her Jim Crow childhood. As civil rights reached rural Kentucky, the freedom to travel fast and stylishly with people of one's own choosing was invested with a strong democratic and counter-cultural charge which brings the transitional phase of teenage high-school desegregation to vivid life: We are looking for ourselves beyond the sign of race. Nature is the only place we can go where race leaves us. We can ride this car way out . . . At first I am not sure I want to cross the tracks to make nice with a white girl but the smell of the car seduces me. Its leather seats, the real wood on the dashboard, the shiny metal so clear it's like glass – like a mirror it dares to move past race to take to the road and find ourselves – find the secret places within where there is no such thing as race (hooks 1997). This history of transgression needs to be acknowledged, but from my perspective it represents the victory of car culture. Instead, I want to claim that those powerful effects must be secondary to our grasp of the destructive and corrosive consequences of automotivity and motorization.

Part of the explanation for these disturbing patterns lies in the fact that African-Americans were in on the automobile revolution from its inception. As the twentieth century unfolded, America's black freedom movements engaged what can be called the poetics of transit in a number of contrasting patterns but the journeys of ex-slaves and their latter-day descendants to the promised land without racial hierarchy were always going to be made on foot or, where that was not practicable, undertaken by freedom train or perhaps by de-segregated bus. 'People get ready, there's a train a comin' and 'Stop That Train' were the epochal cries of Curtis Mayfield and Peter Tosh, may they rest in peace. James Brown's JBs travelled by monorail not limousine. All these locomotives were on an overground railroad: a noisy, visible and powerful symbol that inverted – conjured with – the imperial potency of the industrial technologies to which white supremacy and its signature unfreedoms were articulated. Houston A. Baker jnr. (1984) has unravelled some of the interpretative tangles presented by those 'traditional' railroad tracks, train whistles and crossing points and the world of coerced labour from which they stem. The courageous, volkish preparedness to be even just 'a headlight on a north-bound train' in order to escape from the terroristic, disabling world of Jim Crow can be understood best where his insights have been set to work, but the place of automotive trans-portation in the same semantic and mimetic fields has been seldom approached with comparable rigor and clarity. A very different but complementary shade of blue is described in W.T. Lhamon jnr's brilliant reading of Chuck Berry's 'motorvating' songs, particularly 'Maybellene', an epic narrative of race and gender, and the later 'Promised Land', a complex act of conjuring with the topography of the freedom rides. Lhamon locates Berry and his protagonists in the lineage of the minstrel show but the new era marked by their accelerated insubordination is easy to behold: 'A man who has overcome his victimization by chasing down the fanciest Cadillac on a rainy road, intuitively drawn on the language and riffs of his folkways, intertwined them with that twang and rhythm of his neighbors' counter tradition, and snuck his resulting anthem of pride past all the natural and cultural roadblocks – what might such a sly agent do with straying Maybellene at their peak moment?' (Lhamon 1990).

That image of the roadblock, which Berry shared with Bob Marley among others, compresses the existential perils of driving while black. It is misleading where it suggests the antagonistic relationship between country and city that Berry's drivers were negotiating. Cars belonged first to the city not the woods. The urban environments that they entered and changed were reconstituted around their devastating presence. They have damaged life of black communities everywhere they have helped to produce a new conception of the street as a place of danger and violence rather than community, creativity and mutual-ity.[5] Even that achievement is only a fraction of the automobile's revolutionary significance. Today, their inexplicably neglected history helps to articulate important conceptual and political lessons. It obliges us to update our understanding of culture itself in order to accommodate mobility and speed and their transformation of space, as well as to acknowledge the entrenched divisions between public and private that these ludic technologies promote and invest with new meaning. Along this route, we are also required to adjust the way we understand commodities.[6] These automobiles are products but we have already seen that they can also be conspicuously productive. Their shiny authority and their antisocial prestige alike entail something more than just the terminal point of industrial capital's mystified economic circuits. It bears repetition that cars are integral to the privatization, individualization and emotionalization of consumer society as a whole. They are the ur commodity and as such they not only help to periodize our encounters with capitalism as it moves into and leaves its industrial phase, they also politicize and moralize everyday life in unprecedented configurations. In doing so, the reflexive surfaces they provide show us our distorted selves. Their hold over us also reveals how particular objects and technologies can in effect become active, dynamic social forces. Their power compels an acknowledgement of the conditions under which technological resources can acquire the characteristics of historical agents or social actors. Cars fudge any residual distinctions between base and superstructure by defying the assumed fixity which gave that powerful spatial metaphor its analytic grip. Car culture is therefore the primary challenge to all theories of consumption and postmodernity. Analysing it demands that we modify our understanding of consumer society and its complexities so that it can encompass the alienated but nonetheless popular pleasures of auto-freedom – mobility, power and speed – while appreciating their conspicuous civic, environmental and political costs (see Strasser 1999: 192). Wherever car culture reshapes society, even where consumerism is not the dominant economic pattern, that triumph places technology and power squarely in the middle of ordinary life which is transformed by the ways that cars have redefined movement and extended sensory experience. We need to be able to examine the workings of this ordinary potency within power's wider circuitry: governmental, metropolitan, ecological and, of course, economic. The history of black communities and automotivity can help to accomplish these tasks.

The privatization of mobility and experience associated with these large historical changes is difficult to interpret under the sign of racial difference. The moment of liberation should not be exaggerated, but it should not be ignored either and it would be wrong if the car becomes nothing more than a vehicle for the compromise and accommodation of freshly-tamed black interests to a resurgent consumer capitalism, unconfined by cold-war geo-politics. Once we learn for example, that African-Americans currently spend some 45 billion dollars on cars and related products and services and that 'they are 30% of the automotive buying public although they are only 12% of the US population' it is difficult to resist the idea that the special seductions of car culture have become an important part of what binds the black populations of the overdeveloped countries to the most mainst.eam of dreams. The Motown Corporation's 1975 transcoding of the popular movie 'American Graffiti' into black-alternative version 'Cooley High' could then be used, like their reworking of 'The Wizard of Oz' for African-American children from the same period, to mark the emergence of more extensive operations in the fields of corporate multi-culture, race-branding and ethnic 'cluster' marketing. Perhaps car culture and the unwholesome desires it kindles have become a key component in what keeps the more privileged and wealthy minority groups in the middle of the road? Car culture may be pivotal in foreclosing the possibility of any substantive connections between them and other less fortunate groups, both inside their own society and among the 'third world' folk who live within the veil of scarcity defined, if they are lucky, by the alternative transit order of the bicycle.

This angle of analysis interests me because I am acutely aware that thirty years ago, a radical social ecology that cautiously identified the automobile among the most destructive technological innovations to have been produced on this planet was not alien to the dissenting articulation of black political interests by disenchanted African-Americans.[7] The moral authority of that largely forgotten oppositional legacy must be considered again as a political resource for the future. It raises alternative possibilities and suggests that, even though it may be hard to detect at the moment, there is, dormant in the same black idioms, a critique of car culture which might once again provide vivid ways to make the disquieting perspective of a vernacular, 'green' anti-capitalism intelligible. This tradition of reflection on the wrongs and woes of consumer society has wellsprings in the economic processes that saw black people themselves become commodities for sale on trans-local markets. If those honest pleas to 'down-shift' could somehow be revived, they might be able to bring uncomfortable but unavoidable political and moral choices home to a new generation that has been entirely habituated to the car and its pleasures and which as a result finds alternatives to consumer capitalism exceedingly difficult if not impossible to imagine. The historic role of car dealership in the development of autonomous economic activity and entrepreneurship among black Americans should not be lightly scorned. But it stands in stark opposition to the altogether different tone established by the less materialistic sentiments elaborated, for example, in William De Vaughan's classic song 'Be Thankful For What You Got'. This was a multi-million-selling, mid-1970s recording that remains memorable for Vaughan's repudiation of car culture and its increasingly powerful claims on the racialized ontologies of the ghetto's desperate and immiserated inhabitants. It probably helped that De Vaughan, a guitar-playing Jehovah's Witness from Washington DC, cut this car-sceptical anthem in Philadelphia's Sigma Sound Studios. The Philly house band were at the peak of their prodigious creative capacities and their sinuous playing has sometimes helped to turn his pious attack on car-consciousness into something of a driver's anthem. Nonetheless, the resulting record with its memorable chorus of 'diamond in the back, sun-roof top, digging the scene with a gangsta lean' caught the historical energy of that transitional moment so comprehensively that it drew an interplanetary acknowledgement from George Clinton's Mothership and lives on today. De Vaughan's important tune remains notable for its disquieting, opening lines which offer eloquent testimony to what is now, too often, a heretical possibility:

> You may not drive a great big Cadillac,
> gangsta white walls and a TV antenna in the back,
> though you may not have a car at all,
> remember brothers and sisters, you can still stand tall.[8]

This is a very long way from Missy Elliott's insistent, insinuating request to know 'Beep beep, who got the keys to the Jeep?', a celebration of the experience of rolling with the rag top down that counterposes the 1970s and the 1990s against each other.[9] Notwithstanding Suzanne E. Smith's (1999) arguments for a political and economic linkage between the Detroit car industry and the productive regime established in Berry Gordy's Motown hit factory, it is very odd indeed that the economic strategies currently favoured by America's postmodern black nationalists should be quite so preoccupied with the riches to be won from segregated dealing in automobiles. These, after all, are commodities produced by the very same exploitative corporations that played such an important role in building the racial division of labour which characterized America's industrial capitalism.[10] If this is a lapse, it expresses amnesia-inducing properties of the car which need to be taken into account, particularly if we accept that a critical history of car culture might be implicated in an overdue bid to restore missing elements of black America's collective political memory. Not least of these would be precious knowledge of complex connections between black freedom struggles and organized working-class interests in Detroit and elsewhere; of the ready racialization of Fordism by its footsoldiers and its architects; and of the particular features of racially-divided factory production which need to be placed in relation to the history of slavery if they are to be understood.[11] Nobody can quarrel with the suggestion that Henry Ford was a white supremacist even if his unqualified contempt for the negro was sometimes articulated in paternalistic form. His anti-semitism had been spelled out at an early stage in his pamphlet 'The International Jew: The World's Foremost Problem' which had brought his ideas and achievements to Hitler's attention (see Higham 1983). There is no need to overinterpret the well documented mutual admiration of these two men. In *Mein Kampf* Hitler praised Ford's ability to remain independent against the pressures of international Jewry (1943 [trans. Manheim]: 639) while Ford accepted the Grand Cross of The German Eagle from the Nazi government in 1938. This sort of information, like the news that Ford Werke, the German division of the Ford company, gave Hitler a 35,000 Reichsmark birthday present in 1939 and had deployed the coerced unfree labour of French ,Russians, Ukrainians and Belgians even before the Nazis placed its plant in 'trusteeship' (see Silverstein 2000), should complicate our understanding of the relationship between Fordism and Fascism and invite open-ended questioning of the fact that, at both ends of that historic chain, the car and its automotive freedoms were bound to the private, modern leisure promoted by authoritarian and race-friendly regimes: one corporate, the other governmental.

During the period of decolonization and the era of Black Power that responded to it, the basic idea of black solidarity depended upon building a transnational bridge across the chasm of imperial exploitation and combined but uneven development. The contemporary dominance of car culture is another blunt sign that those valuable historic links are becoming harder to secure. New possibilities for solidarity and translocal action are required now that corporate power can routinely challenge and sometimes even eclipse governmental statecraft. The recent campaign against car culture raised from within the heart of overdevelopment by groups such as Reclaim The Streets has been notably enriched by its connections to the pursuit of rights in areas of the developing world where oil companies have held human life and the biosphere in common contempt. The record of Shell in

Nigeria and Colombia provides some of the best examples here. Oblique connections of this type help to make the car and its cultures into a different kind of issue. The neo-imperial dynamic gets delivered back home, transmitted deep into the tissue of the everyday. Thus the car can help to delineate basic political and economic divisions, particularly those that would sunder the overdeveloped parts of the planet from the rest. Of course, from another angle, we should not ignore the deepening inequalities and persistent poverty that can be found inside the glittering fortifications of car-saturated overdeveloped zones. Similarly, the widening gulf between that world and the systematically underdeveloped spaces on the planet is not visible only in the patterning of transnational relations. The post-colonial metropolis is host to a massive discrepancy in material conditions and life-chances that can be evident even when the groups involved dwell in close proximity. This growing inequality constitutes acute local problems to which resolutely privatized transport often, it would seem, in paramilitary vehicles, presents an attractive solution. The motor car has helped to disseminate and popularize the absolute social segregation that once characterized the colonial city but which is now a widespread feature of postmodern overdevelopment. The car is the primary instrument of what Andre Gorz used to call the 'south africanisation' of social life (Gorz 1989). This means it would be easy to be intimidated by the intractable combination of social, economic, political and cultural problems that the dominance of the private car represents.

Though Europeans had invented the petrol-powered car, it has been an overwhelmingly American technology (see McShane 1994, chapter 6). The initial European dominance of car production was over by 1904 and the quality of US cars improved rapidly. In what was an essentially urban market, Americans were soon far more likely to own cars than their European counterparts. As the century turned, cars had emerged as a critical element in the imperial nation's fantasies of metropolitan order, commerce and reform. The car would improve public health and end the pollution problems caused by horses (ibid.: 122). David Nye, the historian of technology, reminds us that for much of the twentieth century automobile production was the very 'engine of the American economy, stimulating a wide range of subsidiary industries and suppliers' (Nye 1998: 178). Because cars were a US phenomenon, they were comprehensively entangled in that fractured nation's politics of racial hierarchy. Initially, automobiles had been exclusively presented to white consumers. Some companies expressly stipulated that their machines should not be sold even to those few blacks who could afford them. For African-American populations seeking ways out of the lingering shadows of slavery, owning and using automobiles supplied one significant means to measure the distance travelled toward political freedoms and public respect. Employed in this spirit, cars seem to have conferred or rather suggested dimensions of citizenship and status that were blocked by formal politics and violently inhibited by informal codes. Later, the same freedom-seeking people would be confined to the disabling options represented by rural poverty on one side and inner city immiseration on the other. Here again the car provided handy solutions to the eco-political problems it had occasioned and multiplied. Many African-Americans were actively disadvantaged by social and economic changes which made the car into an absolute necessity if employment was to be sought and maintained. Though bell hooks suggested a feminist counternarrative, when the Jim Crow system of segregation was eventually broken, it would be the same motor car that provided both blacks and whites with tacit, complementary solutions to the discomforts of travelling in close proximity with each other. For blacks, driving themselves could be part of liberation from their apartheid, while for whites, this tactic supplied a legitimate means of perpetuating and indeed compounding segregation in the new forms created by essentially privatized transport between suburban dwellings in 'race'-homogenous neighbourhoods and distant employment opportunities. White flight from urban centres was not just accomplished by means of the automobile, it was premised on it.

Needless to say, once they were officially allowed to do so, American blacks bought cars as readily as their economic circumstances permitted. Indeed, because their public displays of wealth had been restricted by both formal and informal rules, cars acquired a special significance in their complex counter-cultures as signs of insubordination, progress and compensatory prestige. The case of Chuck Berry is but one example drawn from their modern traditions in which automobiles were serenaded and savoured in accordance with an aesthetic code that valued movement over fixity and sometimes prized public style over private comfort and security. Motor vehicles were public ciphers of celebrity and black musicians and entertainers in particular appreciated the powerful poetic possibilities that cars created. The paeans they offered to their own transportation, to private car travel, to speed and even to the *idea* of the car itself, might also be used today to construct a shadow history of the shifting appeal of emergent consumer culture to those marginal populations who had not initially been targeted by manufacturers but for whom the dreamworld of American consumer culture held the most potent appeal, precisely because of their exclusion from its land of promise. Here we should recall that the very first verses of that founding text of African-American folklore, the toast, 'Staggolee' refer to the hero's "28 Ford'. Eight years later, Robert Johnson's popular recording 'Terraplane Blues' was one of the first race records to link the motor car to the female body and driving to sex. Johnson's promises to check the oil, flash the lights and get under the hood open up a curious path through the remainder of the twentieth century. It terminates in R. Kelly's 'You Remind Me of Something' a song that plumbs the depths of this estrangement by suggesting that the love of objects, in his case a jeep, now provides a strange template from which the blank dimensions of troubled interpersonal desire are to be shaped. This journey through the psycho-sexual fields of consumer society proceeds by a circuitous route. Albert King's weary complaints against the neo-slave labour that has claimed him on the Cadillac assembly line are only partially redeemed by K.C. Douglas's classic appreciation of the Mercury – a recording memorable for the fact that, like Berry's, it spanned the adjacent but often antagonistic musical worlds of blues and country. On this journey, we overtake Sam Cooke and Miles Davis in their Ferraris and Aretha in her pink Cadillac. Think for a moment about the different 'ethnic' options those brand choices might be used to represent. We press on through Jimi Hendrix's crosstown traffic which articulates more echoes of slavery and presents the woman not as a car herself but as an obstacle blocking the driver's unobstructed access to the city. (It was annexed recently to provide the soundtrack for a Volkswagen advertisement.) Further on up this road, we encounter Prince's little red Corvette and the New Power Generation's 'Deuce and a Quarter'" (a shocking hymn of appreciation to that ghetto chariot, the Buick Electra 225). We are delivered finally to the terminal point provided some sixty years beyond 'Terraplane Blues' by TLC's recent global hit 'No Scrubs' in which the irremediable uselessness of the unworthy black man in question is conveyed and confirmed by the fact that he has no car of his own and has been spotted 'hanging out of the passenger side of his best friend's ride'.

That risky sequence affords important interpretative opportunities. It is a means to observe driving become race-coded service work and then to cease being so. To put it another way, it helps us to measure the gap between being a chauffeur, wanting a chauffeur and having a chauffeur. The chauffeur, a servant who is intimately positioned between the car-owner and the work of driving is paradigmatic of the whole new world of casual and insecure work. The driven are placed in a new position. Their removal from the world of risk is identified as the precondition of their ability to control it. Their sensory apparatus needs to be protected from assault by traffic. The chauffeur exercises and commands the passenger's personal power without owning it. The car's extension of the driver's sensorium so important elsewhere as the index of automotive freedom, is here recognized as an unwanted, dubious privilege. Chauffeur and passenger collaborate in what might be called the 'driving Miss Daisy dyad' a model of interdependency that is best approached as an updated version of Hegel's famous arrested conflict between master and slave. In African-American twentieth-century culture it is of course Richard Wright's Bigger Thomas whose fordist nihilism points toward this possibility. We should remember that his existential tragedy begins with induction into the chauffeur's role:

> The night air had grown warmer. A wind had risen. He lit a cigarette and unlocked the garage the door swung in and again he was surprised and pleased to see the lights spring on automatically. These people's got everything he mused. He examined the car; it was a dark blue Buick, with steel spoke wheels and of a new make. He stepped back from it and looked it over; then he opened the door and looked at the dashboard. He was a little disappointed that the car was not so expensive as he had hoped, but what it lacked in price was more than made up for in color and style (Wright 1940).

Nowadays, the figure of the chauffeur fades away because the car is no longer a place of work. It is almost always a playspace. Albert King's assembly line has slipped out of sight along with the those other topoi of residual slavery the Killing Floor and the Chain Gang. No one in North America's remaining car plants debates C.L.R. James's *Notes On Dialectics* in his or her breaks or strives to apply its insights to the struggle against speed-ups. Workers commute long distances but their driving to their work is not recognized as part of it and is typically presented as a form of recreation. These are the conditions under which the car becomes a comfortable platform for the boomin' on-board sound systems that are celebrated annually in the June issue of *The Source* magazine which includes their car audio guide under the chilling motto 'a ride is a man's best friend'. Auto-motive sound technologies are arrayed alongside the alarms and immobilizers necessary to protect them from the attention of unwanted consumers. The wattage of these mobile sound systems is only half-seriously quantified in the distance measure of audible blocks. 'Let's just say that I spent a nice piece says Usher of his CR210 car stereo with removable face plate and six bass box speakers all by Becker/Porsche, and his two 250 watt Rockford Fosgate Punch amps . . . Usher . . . says he's hitting about a block and a half on the Ghettometer' (*The Source* 1988, June, p. 109). Interestingly, the power to be heard seems to exist quite separately from the prestige and horsepower of the vehicles involved: 'If you were a top music executive, what kind of whip would you be pushin'? A bangin' Benz, A luxurious Lex? How 'bout a thugged out Hummer? Well, no matter what kind of automobile you be drivin', if you really wanted to floss, you'd better have an equally butter sound system to check out all those demos. So for all you car enthusiasts, THE SOURCE went out and picked the brains of the big dogs in the hip-hop industry to find out just how they roll' (*The Source* 1999, June, p. 136). The car emerges from this as a place for listening, an intrepid, scaled-up substitute for the solipsistic world of the personal stereo, a kind of giant armoured bed on wheels that can shout the driver's dwindling claims upon the world into dead public space at ever-increasing volume. We learn for example, that Ronald 'Slim' Williams CEO of Cash Money Records, is the driver of a 1999 Hummer who has had his plexiglass sub-woofer enclosures sand-blasted with the company logo. Slim also explains that his car contains 'Sony Playstation and Nintendo 64, a Clarion VCR and Rosen LCD TV with tuner'. The only thing he thinks is missing is an automatic ignition. 'I want to just push a button and start it without a key; you know?'

Coverage of car culture in *The Source* speaks to the evolution of particular patterns of taste in cars and car paraphernalia. It communicates the consolidation of 'race'-specific markets which have proved extremely important at a number of points in the history of the car industry. Corporate legend has it for example, that a distinct market for Cadillacs among 'wealthy negroes' (at a time when company policy dictated that they should not be sold the cars at all) was what had unexpectedly saved the brand from extinction during the depression of the 1930s (see Drucker 1979 in Rose 1987: 104). Whether or not black buyers actually rescued that particular car company, their appetites for similar upscale products was fed by a harsh social system that prohibited displays of wealth and property by the minority of blacks who were in a position to make these ostentatious gestures. The romance of 'race' and automobility did not end there. These days, *The Source's* coverage addresses all the micro-specific problems involved in 'acccessorizing' one's ride. We're told, for example that 'Rims are what separates the motorists from the enthusiasts. Not only do they exhibit personality, but a fresh set of chrome plated wheels can be the exclamation point of your car's overall look . . . "Right now people want bigger wheels," says Joe Smith of Butler Tires in Atlanta. "High-end customized 18- to 20-inch newer designs."' We learn later that Hill 'has sold to such notables as Erick Sermon and Chili from TLC' (*The Source* 1999, June, p. 144). His hottest and most expensive wheels by Brabus and Lorinser run for $10,000 and $8,000 respectively.

The distinction between motorists and enthusiasts is clearly telling. It suggests first that the need to personalize or 'customize' a car may fall in different patterns among different ethnic or racialized groups, and secondly that subordinate groups might be producing elements of their own understanding of their race-coded particularity by manipulating it. Furthermore, it raises the uncomfortable possibility that, rather than receiving the truths of individual or collective identity from a branded, prestigious or expensive object, these people are projecting their race-thwarted individuality back into the object in ways that reply to the respectable world of official, finished consumerism. By opening up those commodities to ongoing work, by making them into a process rather than a closed artefact, they may on some level be gesturing their anti-discipline to power even as the whirlpool of consumerism sucks them in. It is hard to locate any crumbs of hope in the more pathological results of this process of surrender:

'This is "my big black truck" that I sing about in "This Is How We Do It"' says Montell Jordan as his '95 GMC Suburban accelerates through the slow moving Los Angeles traffic. At the time of his first single, Jordan's road monster was just a fantasy . . . But once that single hit No.1 . . . and once the $64,000 loan he'd taken to attend Pepperdine University was paid off – the truck of his dreams drove into his life. 'Off the lot, the truck only cost forty thousand, tops' explains Jordan, his size 12 Gucci loafer pressing the gas. 'And then another forty to get it customized.' 'Customized' is too tame a word. This vehicle (which is insured for a million bucks) is just plain macked out. There's a TV and VCR with infrared headsets and an onboard video library . . . temperature controlled cup-holders, a mini-humidor stocked with a selection of fine cigars, a digital compass and thermometer mounted on the rearview mirror, leather captain's chairs in front and back, plus invisible speakers cut into soundproof walls. Not all the customizing is for comfort 's sake: Jordan invested in turbocharged engine, Lamborghini brakes, PIAA purple halogen headlamps, and a roof rack for his snowboard . . . any thieves scheming on the GMC jiggmobile are sure to be stymied by a security system that can shut off the engine via satellite (Relic 1988).

Reconstructing the history of these markets and resistance/resignation behaviours takes us back to the usefully monumental figure of the boxer Jack Johnson. He was heavy-weight champion of the world in bold and sometimes reckless defiance of the masculinist codes of white supremacy during the first years of the colour-line century. His well-proportioned ghost remains a significant residual presence even in the different mediascape adumbrated by Michael Jordan's global stardom (LaFeber 1999). Johnson's place in the bitter political conflicts over manliness, masculinity and 'race' that greeted the century of the colour line have been considered at length by other writers but their accounts have not usually been engaged by his inauguration of these distinctive patterns in the way that automobiles were understood and articulated within and sometimes against American racial codes. Johnson's manliness and his ambivalent stardom were both expressed in and through his command of the cars which drew hostility, harassment and introjected, covetous admiration from the police wherever he went. This was partly a reaction to the presence of white women in Johnson's cars but it should also be recognized as involving responses to those cars themselves and to the idea of a black man at the wheel. We miss the significance of Johnson's exemplary life if we emphasize one of these factors at the expense of the others. Though he was among the first African-Americans to fall foul of the law for the informal crime of 'driving while black', Johnson's well-known love of cars and speed enters into his mythification at different points far removed from the contested space of the public thoroughfare. In October 1910, at the height of his visibility, Johnson challenged 'the fastest driver of the day' Barney Oldfield, to a public contest that would inevitably reflect upon the respective capacities of their racial groups in the emergent world of motor-sport. The American Automobile Association threatened to bar Oldfield from their organization if he took part in this unwelcome event which was staged in front of some 5,000 spectators at Sheepshead Bay on Coney Island. Johnson, driving a bright red 90 horsepower Thompson Flyer, was comprehensively defeated by his opponent, at the wheel of a 60 horsepower Knox machine (Roberts 1983). The *New York Sun* greeted Oldfield's victory with the headline 'White Race Saved' (McShane 1994: 134). It is telling that Johnson, an aristocrat of the body who encapsulated the twentieth-century negro's rapturous relationship with physicality, failed to triumph in this novel domain where mental attributes were apparently decisive. The car provided an important new medium for the evaluation of these racialized qualities. The inferiority of the inferior races could now be communicated through the idea that they were bad drivers.

The planetary reach of the African-American vernacular has meant that globalizing black culture has been repeatedly oriented towards north American standards, desires and passions. These inclinations cannot be easily separated from the dynamic, depressing pattern of US race-consciousness and racialized hierarchy. One result of this can be seen in the way that automobiles became more than either simply tools or signs. Once black history and experience were understood and widely experienced as largely urban, if not always metropolitan phenomena, shifting patterns of car ownership, car use and car fascination articulated political and moral problems around the constitution of community and the integrity of its own subaltern public world. The cars and their antisocial sociality redrew the lines between public and private, impacted profoundly upon gender relations, illuminated class divisions, interrogated selfishness, and tested the mutuality of sub-urban residential locations that were increasingly remote from places where work could be found. The institution of American apartheid is closely connected to the extinction of the walking city. This alone suggests that studying car culture can help to clarify the shifting political character of twentieth-century African-American movements toward justice and against the indignities of white supremacy. It is not only as Jack Johnson's example suggests, that car culture provides an important means to comprehend the techno-cultural dynamics of the black masculinity with which it was entangled from an early point. I am certain too that an assessment of the impact of car culture can contribute something distinctive to our understanding of the history of black political activity.

Let us jump-start this concluding part of my argument by introducing another iconic figure. Here is Bob Marley, translocal star on the planetary stage. His was the twentieth century's last effective contribution to the forms of black consciousness that could span the divisions of colonial development and speak in different but complementary ways to dispersed and remote populations, synchronizing their consciousness and their own divergent analyses of their local experiences with a universal language of sufferation. Marley's double consciousness and his subtle grasp of the poetics of Rastafari livity both refined his skill with words. He is notorious for re-naming BMWs 'Bob

Marley Wagons' and thus, from my point of view, confirming the translocal force of the black consumer culture's romances with auto-mobility. However, Marley came from an environment in which, as numerous Reggae tunes testify, the pleasures of speed and automobility were more usually experienced on two rather than four motorized wheels. Long before he capitulated to the allure of the 'bimma', he had been seen driving through Kingston in a second-hand Hillman Minx bought in 1968. Still less familiar is the time that Marley, who was after all the son of a green card Jamerican and had ducked his call-up papers for the Vietnam war by fleeing back to the Caribbean, spent working in the US automobile industry producing cars in the Chrysler plant in Wilmington, Delaware. The pain of that experience is most obviously audible in the song 'Night Shift' but it is not only there. Like the time he spent working as a janitor in Wilmington's Dupont hotel, it clearly conditions other aspects of his moral and political distaste for the corrupting superficiality of Babylon queendom. I want to end with that portrait of Marley not as an anxious, upwardly mobile driver cocooned in a shiny private shell of German techno-aesthetic excellence 'accesorized' with his vanity plates, but as a grumbling and disaffected worker in the car industry. It was there, in that crucible, on the seat of that forklift truck, that his acute understanding of the relationship between slavery and wage slavery was refined. Perhaps from those moments of disenchantment a twenty-first century critique of consumer capitalism might be reintroduced into the vacuum that black political thought has become?

References

Adorno, T.W. (1974), 'More haste, less speed'. In *Minima Moralia,* trans E.F.N. Jephcott, London: Verso.

Baker, H.A., jnr (1984), *Blues, Ideology, and Afro-American Literature: A Vernacular Theory.* Chicago: University of Chicago Press.

Baudrillard, J. (1981), *For a Critique of the Political Economy of the Sign.* trans C. Levin. St. Louis: Telos Press.

Drucker, P. (1979), *Adventures of a Bystander.* New York: Harper and Row.

Dusenberry, J.S. (1949), *Income, Saving and the Theory of Consumer Behavior.* Cambridge, Mass.: Harvard University Press.

Flink, J.J. (1990), *The Automobile Age.* Cambridge, Mass.: MIT Press.

Gorz, A. (1989), *Critique of Economic Reason.* trans G. Handyside and C. Turner. London: Verso.

Higham, C. (1983), *Trading with the Enemy: an Exposé of the Nazi-American Money Plot, 1933–1949*. [placename]: Delacorte Press.

Hitler, A. (1943), *Mein Kampf.* trans R. Manheim. Boston: Houghton Mifflin.

hooks, b. (1998), *Wounds of Passion*. London: The Women's Press.

Kay, J.H. (1997), *Asphalt Nation: How the Automobile Took Over Ameica, and How we can Take it Back*. Berkeley: University of California Press.

LaFeber, W. (1999), *Michael Jordan and the New Global Capitalism*. New York: Norton.

Lhamon, W.T. jnr (1990), *Deliberate Speed: The Origins of a Cultural Style in the American 1950s*. Washington, DC: Smithsonian Institution Press.

McShane, C. (1994), *Down the Asphalt Path: the Automobile and the American City*. New York: Columbia University Press.

Meier, A. and Rudwick, E. (1979), *Black Detroit and the rise of the UAW*. New York: Oxford University Press.

Molnar, V. and Lamont, M. (forthcoming), 'Social categorization and group identification: how African-Americans shape their collective identity through consumption'. In K. Green et. al., *Interdisciplinary Approaches to Demand and its Role in Innovation*. Manchester: Manchester University Press.

Nadis, S. and MacKenzie, J. (1993), *Car Trouble*. Boston: Beacon.

Nye, D.E. (1998), *Consuming Power: A Social History of American Energies*. Cambridge, Mass.: MIT Press.

Relic, P. (1988), 'Big Black Truck: buckle up for an expedition in montell jordan's jiggmobile'. *Vibe*, June/July, p. 117.

Roberts, R. (1983), *Papa Jack: Jack Johnson and the Era of White Hopes*. London: Robson.

Rose, D. (1987), *Black American Street Life: South Philadelphia 1969–1971*. Philadelphia: University of Pennsylvania Press.

Ross, K. (1995), *Fast Cars, Clean Bodies: Decolonization and the Reordering of French Culture*. Cambridge Mass.: MIT Press.

Silverstein, K. (2000), 'Ford and the Führer: new documents reveal close ties between Dearborn and the Nazis'. *The Nation*, 270, 3, 11–18.

Smith, S.E. (1999), *Dancing in the Street: Motown and the Cultural Politics of Detroit*. Cambridge, Mass.: Harvard University Press.

Smith, T. (1993), *Making the Modern: Industry, Art and Design in America*. Chicago: University of Chicago Press.

The Source (1988), June.

The Source (1999), June.

Strasser, S. (1999), *Waste and Want: A Social History of Trash*. New York: Metropolitan Books.

Weems, R. (1994), 'The revolution will be marketed: American corporations and black consumers during the 1960s, *Radical History Review,* 94–107.

Weiss, M.J. (2000), *The Clustered World.* Boston: Little, Brown.

West, C. (1993), *Race Matters.* Boston: Beacon.

Wright, R. (1940), *Native Son.* New York: Harpers.

Notes

1. I must thank Radiclani Clytus for the information that this seems to be an abbreviation of the phrases 'whip n' ride' or 'whippin' ride'.

2. 'When Buick wanted to increase sales of its luxury Park Avenue model to minority buyers, it noticed that African Americans frequently have lower rates of home ownership but more disposable cash than whites at similar income levels. Affluent African Americans, the company discovered, were being ignored as potential Park Avenue purchasers simply because their status as renters didn't fit the traditional buyer profile ... "a lot of companies pay lip service to targeting minority groups" says Patrick Harrison, marketing line manager at Buick. "But we wanted our Buick dealers to develop a personal relationship with African Americans. And this is how we found a way to bringing minorities into our dealerships"' (Weiss 2000: 96–7).

3. See also the car pages of Melanet which can be viewed at melanet.com.

4. Black America's distinguished public intellectual Cornel West concedes this much in his book *Race Matters* (1993): he describes leaving his 'rather elegant' car in a safe parking lot before riding uptown to East Harlem to have his photograph taken. The desire for unobstructed access to Manhattan taxis is a recurrent feature of much of the discourse recently produced to communicate the hurt and injury of US racism. The full ground-shaking force of the primal scene in which taxi travel is denied by the leviathan of white supremacy can only be appreciated, I suggest, in the context of the romance of automotivity. The idea that public transport might be an alternative never comes into consideration.

5. The 'alley' is for example no longer a place of play and self-discovery.

6. Kristin Ross approaches the same abyss when she writes '... the car *is* the commodity form as such in the twentieth century, an argument that becomes all the more convincing when we remember that "Taylorization" – the assembly line, vertical integration or production, the interchangeability

of workers, the standardization of tools and materials – "Taylorization" was developed *in the process of producing* the "car for the masses" and not the inverse' (1995: 19). See also T. Smith 1993. esp. part 1.

7. Think for example of the lyrics of Marvin Gaye's 'Mercy Mercy Me (The Ecology)' from the album *What's Going On?* or Curtis Mayfield's 'Future Shock' from the album *Back To The World*.

8. William DeVaughan, 'Be Thankful For What You Got', Roxbury Records, 1975.

9. Missy Elliott, 'The Rain (Supa Dupa Fly)', East West 851, 1997.

10. See for example www.aacgg.com/ the African-American car-buyers' guide that has been put together under the aegis of 'Melanet' the 'uncut black experience on the internet'. See also Meier and Rudwick 1979 and Flink 1990: esp. 126–8.

11. James Flink who has coined the useful concept of 'Automobility' addresses these issues with great insight. 'In the black ghetto of Inkster, Michigan, adjacent to lily-white Dearborn, a bastion of the Ku Klux Klan, Ford set up a gigantic plantation for his black workers. He paid them only $1 a day in cash of their $4 wage, the remaining $3 in food and clothing from a public commissary. Seeds to plant in garden patches and communal sowing machines were also furnished to Ford's Inkster blacks' (1990: 221).

Raggare and the Panic of Mobility: Modernity and Everyday Life in Sweden

Tom O'Dell

Now everything arrives without any need to depart (Virilio 2000: 20)

In Sweden, mass-consumption and 'America' are two concepts which are often understood in terms of one another. Ask what is American, and Swedes will draw up a list of consumer items which are commonly found throughout the country and which are used daily without the slightest reflection. In fact, while Swedes commonly assert that 'Sweden is the most Americanized country in Europe', they tend to conceptualize this is in terms of, and through the act of, consumption. Thus, activities such as purchasing frozen dinners, wearing athletic shoes, and watching too much television are not just understood as forms of consumption in Sweden: these are all activities which are generally interpreted as the most concrete evidence of a particular form of transnational movement popularly encrypted as 'the Americanization of Swedish culture'. This being so, it should also be pointed out that to the extent that Sweden has been Americanized, it has been Americanized in a very Swedish way: through the interaction of Swedes with each other, via their own consumption practices, and within the context of Swedish history.

This chapter takes the American/Swedish juncture as a point of departure for an investigation of some of the workings of the transnational. Using the American car as an example, the text proceeds by examining consumption and mobility as the idioms through which transnational cultural elements are implemented in the Swedish setting to express alternate gendered conceptions of identity. These identities are themselves considered in terms of the aesthetics of everyday life, whereupon modernity is identified as a key metaphor used in both Sweden and the United States to delineate middle-class values. It is here argued that far from being a uniform global process, modernity is a multifaceted and polyphonic phenomenon whose effects can be quite different from one cultural setting to the next. Finally, the analysis concludes by drawing attention to the manner in which working-class identities based around the American car are simultaneously produced in connection and confrontation with multiple localities ranging from specific local settings, and national institutions, to transnational media flows.

The Aesthetics of Swedish Modernity

Before engaging the American car, and the people around it, it is necessary to consider briefly the context within which the following discussion is set: the post-war Swedish setting. In so doing, however, I do not wish to exaggerate the uniqueness of the Swedish context. In many ways it is perhaps quite similar to many other European and American settings. Despite this, however, the following discussion is based, in some ways, on the understanding that everyday life in Sweden is not exactly like everyday life in the United States or most other European countries. It is similar, but also different, and it is this tension which saturates most of the following discussion.[1]

Among other things, for example, it has been argued that social mobility is one of the prominent characteristics of post-war Swedish society. Many Swedes have made the journey from one class to another: from working class to middle class, from middle class to upper-middle class and so on. In line with this, Jonas Frykman has argued that as a result of this class-travelling, the Swedish elite stratum has become a heterogeneous group with origins in a large variety of social backgrounds. As individuals they have tended to tone down their unique backgrounds in an attempt to melt into the background of their new class. In this context, Frykman goes on to argue, attributes which are stereotypically considered to be Swedish such as rationality, shyness, and 'grayness' may be seen to have utility in particular Swedish settings, and to this extent, 'many "typically Swedish" characteristics have been useful for those coming from a humble background into a qualified profession. They have had to refrain from articulating their own style, displaying their own pattern of culture. Being gray was a strategic asset here' (Frykman 1989: 43).

However, if the business of everyday life has favoured the maintenance of a reserved disposition, then it might be argued that the privacy of the home has provided an opening for greater experimentation, and this in turn may be one of the reasons that the home has become such an important point of departure for everyday life in post-World War II Sweden, as in other Scandinavian countries (cf. Gullestad 1991: 488). It was here, in the confines of the home, somewhat protected from the peering eyes of colleagues and mistrusted neighbours, that Swedes were able to experiment with new ideas, designs and identities without risking immediate disapproval. Indeed, some have argued that Swedish modernity gained an important foothold in society through its institutionalization as a private consumption project to the extent that Swedes became modern by making their homes modern, and came to measure their own modernity in comparison to their neighbours' in terms of refrigerators, telephones, and television sets (cf. Löfgren 1990: 25, 1991: 111).

At the same time, however, Swedish modernity in the post-World War II era was not defined as consumption for consumption's sake. In Sweden, as in many other Western European countries, *modernity reflected and implied a belief in science, practicality, reason, and rationality* (cf. Frykman 1981; Löfgren 1993; O'Dell 1998). This was the realm and age in which scientists and experts were fetishized and held up as almost God-like figures who could eradicate any of society's problems through their various expertise, making life better in a single stroke through the production of new knowledge and technology (cf. Frykman and Löfgren 1985; Broberg and Tydén 1991). It was an aesthetic and ideal which penetrated even the most trivial and private spheres of daily life. The purchase of aerodynamically-shaped refrigerators, vacuum cleaners, toasters and telephones was consequently legitimated not because it was claimed that they were fun or exciting, but because the purchase of these goods could be justified as a means of *rationalizing* household chores, and making everyday life more practical. In this sense it can be argued that when Swedes entered the Post-World War II age of mass-consumption, they did so under the guise of modernity. This, however, is not to say that Swedish everyday life became gray or monotone in any way; on the contrary, while the streamlined forms of these new modern objects may have been justified through an appeal to their practicality, they also sparked the imagination and tantalized the fantasy. They were excitingly new and seemed to be the material evidence of the promises of science. Embodied in these objects and their aerodynamic forms were small details whose *aesthetics* symbolically spoke of their owner's focus upon the future, and a rational

no-nonsense determination to reach that future as quickly and effectively as possible.

Additionally, but in conjunction with this, being modern also
necessitated an awareness of events and inventions, from family-
planning to streamlined refrigerators, found *beyond* the nation's borders.
Being modern was in this sense the antithesis of being provincial, which
was itself a chronic worry and problem for Swedes (cf. O'Dell 1997:
121 ff.). However, while it was important to demonstrate a sophisticated
awareness of the international, it was just as important that international impulses when incorporated into Swedish everyday life be
subordinated to the middle class's definition of 'good taste', which was
itself closely defined in terms of the functionally subdued and the
soberly rational. As Orvar Löfgren has pointed out, this was the time
in which:

> The luxurious vulgarity typical of German, English, or American family
> rooms, with all of their frilly and superfluous large flowery patterns and
> stark colours were put on the run in Sweden by tranquil pine surfaces,
> small checkered patterns, pale pastel colours, and all that was close to
> nature (1991: 113).[2]

This description is not significantly different from that presented to
me by Åsa, one of my informants who grew up in the 1950s and 1960s:

> You know what? When I grew up you were supposed to have gray, or
> brown, or blue, or beige, or colours like that. And there were always
> people telling you that you couldn't buy a red cape because then people
> would say, 'Look there, it's the girl in the red cape'. . . You were always
> supposed to be dressed in discreet clothing and be made-up discreetly
> . . . because if you did (not), then you were considered vulgar.

Within this context, flashy clothing, colours, and consumer goods lay
beyond the realms of the aesthetics of Swedish modernity. Far from
being perceived in terms of practicality, such items and styles were
dismissed by the middle class (and much of the working class) as
narcissistic attention-grabbers, bordering on the brink of the abnormal,
if not transgressing this line entirely. The result here was that as the
aesthetics of Swedish modern living became ever more clearly defined
and incorporated into everyday life, the boundaries delineating normal-
ity became increasingly vivid. In the end, the trick for the modern
Swede was to partake of the international and demonstrate a sensitive

awareness of it, while simultaneously not breaking with the aesthetics of Swedish modernity.

This being said, it should be pointed out that, within the anthropological literature, there has been a tendency to end the analysis at this point, and thereby underscore the authoritative voice of Swedish modernity by continually describing everyday life in Sweden – and indeed Swedishness itself – through an appeal to the metaphors of rationality, functionality, and practicality.[3] In the remainder of this chapter, with the aid of the automobile as an example, my intention is to add further nuance to this rather homogeneous image of Swedish modernity. As I have argued above, the aesthetics of Swedish modernity constituted an extremely important point of orientation for daily life throughout the country, but the voice of this modernity did not go completely unchallenged. Swedish everyday life was not completely insulated from the outside world. To the contrary it was intimately interconnected with the realm of the transnational and, as shall be argued, these interconnections provided the impetus for new transformative processes of hybridity in the local setting. And in this context, America proved to be a powerful source of change and affectation as the car developed into a catalytic arena around which these processes took place.

The Rise and Decline of Chromed Dreams

In the immediate post-World War II era, America symbolized for many within the Swedish middle and working classes an economic and productive force which could not only rebuild Europe but also serve as a model of industrial and democratic efficiency in Sweden. Such ideas were supported in part by industrial kingpins who increasingly looked west for new and 'better' management strategies, but they also grew out of popular understandings in which America symbolized the full potential of consumption dreams and ideals of freedom (Löfgren 1988: 192). From the world of business, to the family project of decorating the home, America exemplifed efficiency, rationality, and the land of tomorrow (Löfgren 1990; Schoug 1991). In this context, the Americanness of the American car made it special. Johan, a sixty-year-old employee at GM Nordiska AB, summed up the attitude of the time in these terms:

> America and the American people had such an unbelievably good reputation here in Sweden. And besides that, people thought that these cars were so incredibly beautiful. It was art work actually, if you look at

the front of an Oldsmobile ... They were enormous. And people probably also began to dream about the future which would come a bit later. When everyone, almost everyone, would be able to buy their own car. But things were just in the dream stages then, and highest up at the top of the dream pyramid, there was an American car (Johan 60, employee at GM Nordiska AB).

Figure 5.1 In the Power Big Meet in Västerås (outside of Stockholm) thousands of American cars are lined up and displayed. Finns and chrome are among the more popular attractions, Photo: Tom O'Dell.

In the world of Swedish working- and middle-class dreams, the American car had a privileged place; it was the ideal which represented the beauty and potential of things to come. The car concretely symbolized the better standard of living which the Social Democrats had promised the welfare state would provide. It was the material standard which was increasingly available to the common person, and as such, the beauty of the car's chromed grille lie in the fact that it foreboded the seemingly inevitable convergence of a dream world, and the world of actual everyday life.

However, as the 1950s came to a close, American cars increasingly came to resemble rockets and airplanes more than they did automobiles. As a result, it became increasingly difficult for Swedes to conceptualize

American cars in purely functionalist terms, and sales of the cars plum-
meted. In 1951, for example, 8.4 per cent of the new cars sold in Sweden
were produced in the United States. By 1959, when the styling and
size of American cars was at its most extreme, only 1.6 per cent of the
new cars sold in Sweden were produced in the United States. In terms
of sheer quantities, the number of American-produced cars sold in
Sweden dropped from 5,217 in 1951 to 2,676 in 1959, at the same time
as the total volume of new cars sold increased from 62,503 to 169,014.[4]

Part of the problem was that American cars were built and designed
to accommodate the aesthetics of middle-class America's vision of
modernity.[5] They were designed to appeal to American visions of
modernity, built upon the themes of power, speed, and 'the desire of
Americans, as individuals, to participate in the technological revolution . . .' (Hine 1986:90). In line with this, the car was designed to
glide forward unhindered and unaffected by any turbulence. American
modernity implied movement; it implied getting ahead and making
something of oneself. Upward social mobility was the measure of
success for the modern middle-class American, and it was commonly
believed that this could easily be measured in economic terms. It was
this ideal of unimpeded forward movement and space-age technology
which the American car makers used to allure buyers to their cars:

> Looks can be deceiving. It's true, too. Take the '59 Oldsmobile. Expensive-
> *looking*? Yes. Expensive? Absolutely not! In fact, your Oldsmobile Dealer
> can *prove* that it costs less than you'd ever guess to own and drive an
> Oldsmobile Dynamic 88! See him soon – your Oldsmobile Quality Dealer
> – celebrate Springtime in a Rocket – a Rocket that's priced just right for
> you. And remember – proven *quality* is standard equipment on every
> Oldsmobile (30-Second American Radio Commercial, April 1959).

The references to the Rocket (which was actually the name of an
Oldsmobile engine) and Springtime played upon the thought of an
accelerated trajectory into a better and more prosperous future. The
thought of celebrating 'Springtime in a Rocket' undoubtedly raised hairs
of excitement on the back of the necks of much of the white middle
class, not only because the metaphor implied the possibility of embark-
ing upon a new adventure, but also because this ever-accelerating
individual trajectory was and is the driving motor of American modernity. The car symbolized how far one had come on the road to success.

The commercial also prioritizes *looks* and the importance of *looking*
as though one possessed wealth. What was (and still is) important from

the American middle-class perspective was the position which one occupied in society in the present – and the potential trajectory which that position implied for one's future life path. The American middle-class strategy was *not* to conceal a humbler background, as might have been the case in Sweden, but to mark as clearly as possible where one currently existed on the road to well-being. The American emphasis was upon looking as though one were in control of the situation, and appearing as though one had the power and technological gadgetry to make the best of every situation. Driving a car which 'looked' expensive was important, because when one was otherwise an anonymous face in the crowd, looks meant everything. In America, anonymity was more of a problem to be overcome than the strategy of choice as Frykman implies was the case in Sweden (1989).

Juncture and Disjuncture in the Transnational Flow of Culture

However, this Oldsmobile Rocket advertisement does not just exemplify the American aesthetics of modernity. The advertisement is also significant because it was written in the United States by Americans and sent to Sweden for use, and thus provides a potential point of departure for a focused study of transnational cultural processes. This example is interesting because the predominant American aesthetics (flaunt your position so others know where you stand) starkly contrast with the dominant Swedish aesthetics (don't reveal where you come from and keep a relatively low profile), and thus one would expect that any American influences in the Swedish context would be relatively easy to identify.

The result? Unfortunately, the American commercial was irrelevant in the Swedish context due to Swedish laws prohibiting all radio commercials. The advertisement and its message completely disappeared, so it is impossible to guess how it might have been received in Sweden. However, a year later the Swedish promoters at GM Nordiska AB came out with a promotion campaign of their own, and its message was strikingly different than that of the original Rocket advertisement.

In 1960, *GM Revyn*[6] printed an article entitled, 'The peep box became a part of the Folk Home: Meeting with Lena Larsson'. In this article, Lena Larsson, an influential interior designer who has helped define middle-class Swedish aesthetics since the 1940s explained how a family room should be organized in the ideal home. She argued for light, spaciousness, and the construction of a room in which one could be active. The objects in the room should be durable and practical, although not too abundant. 'Empty space is actually the most important thing to have when you furnish a room' (page 13).

General Motors was quick to juxtapose its cars with Larsson's aesthetic ideals:

> Light, space, sure style, and a reflection over functionality characterize Lena Larsson's creativity. In the series of the newest GM cars you'll find some objects which are good exponents of what this implies. And the connection with cars is perhaps not as far-fetched as you would think: we want to have practical, durable and well furnished cars, we ought to have such homes as well (page 17)!

The Swedish middle class itself complained that American cars had become too big; they used too much fuel, had too much chrome, and had fins which were too large. They were just too ostentatious. However, the manner in which the cars were presented here can be seen as an attempt to appeal to Swedish middle-class aesthetics. The cars are described as spacious rather than large. They have large 'panorama' windows and are therefore light. The promoters even throw in the key word of 'functionality' in their juxtapositioning of Larsson's aesthetics and the American cars, although it is unclear as to what it is that is 'functional' about these cars. And above all else, the cars are said to be practical and durable. There is no mention of Rocket motors, or expensive looks as we find in the American radio commercial.

Ironically, when *GM Revyn* pubished an article two years earlier in which it asked what women looked for in a car, the response it received seemed strangely out-of-step with the direction in which GM's stylists were taking the car:

> Not too much glitter, it shouldn't be, so to say, too loaded with strass, but should rather have calm beautiful, and preferably long lines. No sharp edges, a 'beautiful finish' – and sober colour. Who hasn't dreamed of a white car with red interior . . . (Fall 1958: 3)?

In short, the American stylists at GM moved in their own direction, and Swedes, unable to find what they were looking for at GM, increasingly turned to either more sober European products, or alternatively, the simpler American models with their smaller six-cylinder engines and stripped-down ornamentation (Eriksson et al. 1959, 16: 27).

The problem lay in the fact that American aesthetics were 'display' oriented – as a result, the ornamentation of American cars became increasingly extreme – and while Swedes also displayed their socio-economic success, they chose to do so in a more subtle manner, purchasing cars such as Mercedes which may have been expensive, but which did not attract attention like the large chromed American cars. The American aesthetics of 'self-assertion' were not compatible with the Swedish aesthetics of modernity in which 'the beautiful was always the same as the utilitarian, the simple, the practical, the restrained' (Löfgren 1993: 10). Seen in this light, the American car was all but beautiful. The Swede who bought an American car ran the risk of being accused of being irrational, pretentious, or having 'bad taste'. As a consequence, Oldsmobile sold only 32 new automobiles in Sweden in 1960, while Mercedes sold 6,320 and Volkswagen sold 30,791.

Despite this, the life history of the American car did not end in the late 1950s; from a middle-class perspective, the car became something of a vulgarity, and it was just the vulgarity of the car which gave it a new life – this time in the hands of young working-class males called *raggare* (greasers). *Raggare* invested their money in run-down inexpensive American cars which they then decorated outlandishly. At first this implied settling for older cars from the early 1950s. However, by the late 1960s, the larger chromed American cars from the late 1950s had depreciated sufficiently in value to be affordable for working-class youths. Ultimately, it was just these cars which the middle class had by and large rejected, that became (and have continued to be) the cars of choice among *raggare*.

The primary reason for this choice lies in the fact that these late 1950s models, and especially those from 1959, *looked* very different from the other cars on the road, and it was here in the alterity of their aesthetics that a large part of their attraction lie for *raggare*. When economic constraints prevented these young men from obtaining such cars, their second choice was more often than not directed towards older models sporting two-tone paints jobs and as much chrome as possible. The cars became a forum for self-expression, and *raggare* developed their own aesthetic code which was at least partially a reaction against the dominant and normative Swedish preference for the practical and rational. They dressed their cars in blinking Christmas trees, Davy Crocket racoon tails, and pink leopardskin seat covers. They became notorious for cruising slowly and provocatively around town, hanging out of the windows of their cars, screaming, and trying to pick up young women (Friman et al. 1991; *Raggarungdom* 1962: 10). And in the 1970s, when the Swedish middle class was most actively protesting the Vietnam war, *raggare* made a point of waving and prominently displaying American flags in their cars, while driving alongside and heckling marching anti-war demonstrators. The cars were impressive and thought to *look good*, and the goal for these youths was to be *seen* driving in style. However, in order to achieve this end, it was important for youths to personalize the cars and make them their own.

In addition to the aesthetic impression the car created, there were other issues which made it seem increasingly provocative and vulgar in the eyes of the potential middle-class consumer. In hindsight one can say that the social history of the automobile was on the verge of veering off in a new direction; at the time, however, people seemed simply to be unsettled by the fact that the car was, for the first time ever, coming into the hands of working-class youths. And when given the choice, these youths chose American-made models over their European rivals. In retrospect, this is perhaps not so surprising; working-class youths were also captivated by the images of America which had gained currency in Sweden. Freedom, (consumer) democracy, and the promise of tomorrow were all appealing qualities to these youths who also strove after upward social mobility and the creation of their own distinctive identities, and in this sense, their own ideals were highly congruous with those of the majority of other Swedes. However, once in the hands of youths, the car became a threat to the middle class, and an object of debate – and at the heart of the debate lay concerns about morality, mobility, violence and sexuality.

In reaction to a steady stream of complaints by city residents, a number of critical newspaper articles in the Stockholm papers (Rosengren 1997: 150ff.), and rumours that alleged that these car enthusiasts were 'picking-up' and having sexual relations with twelve- to fourteen-year-old girls, the Planning Board of Greater Stockholm, in February 1962, appointed a special commission to investigate the phenomenon of youths and cars. A half-year later, the commission presented a thirty-page analysis – full of statistics, summaries of police reports, and the concerns of the child welfare committee – discussing the social problem which had come to be known as *raggare*. In opening, the commission could report that:

> The word *raggare* as a term for young motorists with certain common features in regards to their mechanical equipment and behaviour, began to appear in the press some time around 1957 or 1958. What was characteristic of these youths was from the beginning that they rode around in large, older-model, American cars, often painted and decorated

in an attention-grabbing way. They hung around in large groups along the main thoroughfare in the centre of Stockholm and at a couple of coffee houses in the city suburbs. Some of them were united in clubs with more or less imaginative names (*Raggarungdom* [Raggare youth] 1962: 6).

Figure 5.2 Caught up in the exhilaration of the moment, six young men arrive at a car meet and prepare for a weekend of adventure in the summer of 1995. Photo: Tom O'Dell.

And while the report went on to assert that 'the stories about *raggares'* ravagings, which have dwelled upon the promiscuity of 12- to 14-year-old girls, are in all likelihood essentially exaggerated' (1962: 28), the general public remained less than sure that they could trust mobile groups of youths bearing such names as Car Angels, Road Devils, and Teddy Boys (all of which were in English).

Fanning the fires of anxiety, the mass media offered an array of accounts, the tone of which was often highly dramatic:

It's late, the country road lies empty and desolate for many miles ahead. The forest stands dark and tight. The beam of the searchlight slices through the night, sweeping, searching. There . . .

Two girls emerge from the side of the road. They are freezing in the winter chill, blinded by the strong light. Inside the car it's warm, and there's German tango music. And two strangers in leather jackets. The girls have waited to be picked up. A minute later the road is empty again. A black *raggarbil*[7] has merely disappeared into the darkness and silence (*Bildjournalen*[8] 1959a: 3).

Another popular tale from the era focused more upon the urban experience:

A girl in jeans, sport jacket, and shawl . . . is hanging around in front of store display window on Kungsgatan [a central street in Stockholm, author's note]. A fancy car pulls up to the curb and a guy hanging out the window whistles. 'Do you want to come along?'

She exchanges a few words with him and climbs into the car. There are already a few other young men and women sitting inside. The car drives away.

'I can't believe that the police don't do anything,' says an upset woman who is waiting for the last bus home to Kungsholmen (*Bildjournalen* 1959c: 9).

The sexual threat which the American car represented for working-class and middle-class parents cannot be overestimated. There were few fates which could be more anxiety-provoking than the thought that one's daughter was spending time riding around in a *raggarbil*. For even if the particular woman involved did not consent to the advances of these leather-jacketed young men – who commonly referred to women as 'meat' or 'lamb's meat' – there were plenty of stories circulating about girls who were raped, or dumped out of the car, miles from town, when they did not give in to the *raggare*'s demands (see *Se* 1961: 40ff.; Spångberg 1962). These fears culminated in 1963 when a special law was passed, popularly referred to as the '*raggare* paragraph', which gave police the right to forcefully remove young women from the cars and company of *raggare* if there was any evidence that their health might be at risk, or that their actions could lead to asocial behaviour.

Issues of sexuality, allusions to prostitution, and fears about the moral delinquency of youths in general were all heavily laden in the narratives of the time (and were to a great extent the primary factors which catalysed reactions from parents, authorities, and the mass media). However, in order to understand fully the moral indignation generated by the situation, it is necessary to reflect more closely upon the role that gender, mobility, and travel played as central metaphors in these tales.

Considering the issue of travel, a number of cultural theorists have drawn attention to the fact that travel and mobility have long been framed and understood as very masculine activities, often framed in terms of exploration, conquest and adventure (cf. Clifford 1997: 31ff.; O'Dell 1999a: 24ff.; Rojek and Urry 1997: 16). Janet Wolff takes this line of reasoning a step further and notes:

> The gendering of travel is not premised on any simple notion of public and private spheres – a categorization that feminist historians have shown was in any case more an ideology of place than the reality of the social world. What is in operation here *is* that ideology. The ideological construction of 'woman's place' works to render invisible, problematic and, in some cases impossible, women 'out of place' (1995: 127).

In this regard, the women who associated with *raggare* were 'out of place' to the extent that they were no longer in the home. At the same time, the 1950s were a period in Sweden in which women left the home for the workplace in increasing numbers. The activities of the young women around the American car reflected in this sense the larger cultural phenomenon of women rejecting traditionally bound domestic roles (Bjurström 1990: 216). However, the public debates about *raggare* took this issue a step further. The problem was not simply that *raggare* and the American car lured some young women out of the home, but that they lured these women, at least temporarily, out of *place* entirely. That is, stories like the ones presented above were not in the first case about women who left home. They were about women who 'disappear', and cars which 'drive away' (with women in them). In addition, the women portrayed in these narratives are not victims (at least not yet), but to the contrary, they are described as willing and active participants who have 'waited to be picked up' and voluntarily 'climb into the car' (see the quotes above). Seen from this perspective, one of the central problems activated by the conjunction of women and the American car was not just that women were 'out of place' but that they really were not in any place. They were in motion, and they *actively* chose to be in motion.

In order to deal with this, a series of actors and institutions were activated to counter this new-found mobility. The '*raggare* paragraph' was one of the instruments designed to achieve this goal. Its intention was to 'save' young women from the moral, cultural, and physical consequences of mobility, but it seems, at least in some cases, to have contributed to the sense of adventure surrounding the car for young women. In line with this, Annette, a woman who lived in a small town in Southern Central Sweden, and enjoyed being in the company of *raggare* and their cars in the mid-1960s, recalled the following experience:

> When you were out in the city, the police could tell you to go home; you were known. You didn't dare to do anything else. It was the same if the social welfare officers said anything to you. Sometimes my brother was out looking for me, but my friends warned me so I avoided him (Quoted in Turesson 1987: 33).

In part, her words reflect the degree to which life around the American car could take the form of something akin to a cat-and-mouse game: an adventure in new forms of rebellion and partial freedom from the domestic realm. However, they also reflect a more serious side of the situation. She was 'known'. The police and social workers had their eyes on Annette and women like her. While things have gone generally well for Annette, such was not the case for all of her peers. In fact, at the extreme end of the scale, the consequences of being associated with this particular group of youths and the American car could be horrific, as illustrated by the following passage taken from a doctor's journal in 1960:

> In the course of conversation she demonstrates her lack of judgement in many ways. Speaks with pride about the fact that she is a *raggarbrud*[9] and even speaks unrestrained and openly about her life among *raggare*. Completely lacks ethical–moral conceptions. Says she does not like *raggare* entirely because the have to obey the law of the *raggare*. Probably has sexual intercourse as a *raggarbrud* to the extent that the law of the *raggare* allows and to the extent that they demand it of her. Does not seem to possess any form of self-restraint whatsoever.
>
> In terms of intellect, she seems impoverished, and can without a doubt be regarded as an imbecile, even if she belongs to the upper stratum of this category . . . She cannot manage a life in freedom at this point in time; indeed to the contrary lives in complete promiscuity.
>
> . . . It is inconceivable that this woman will ever be in such a psychological condition that she can take care of a child. Since she obviously has perpetual random sexual relations, there is naturally a great risk that she will become pregnant again. I am, therefore, of the opinion that sterilization is completely in order (Broberg and Tydén 1991: 128).[10]

The doctor's 'opinion' concerning the best way to treat this particular seventeen-year-old girl was all-important. The issue was forced sterilization, a reality in Sweden from 1935 to 1975, and the 'patient's' own thoughts about the suggested 'cure' were of little interest or consequence in these cases.[11] As a means of legitimating the opinion that the young woman should be sterilized, the doctor supports his conclusion by referring to the woman's supposed low intelligence, and her presumed excessive sexual activity. However, one can question the degree to which these were actually the critical factors behind the final recommendation. There were, after all, plenty of people throughout Sweden who could have fallen into either (or both) of these categories, but who were never in jeopardy of being subjected to this treatment. Decisive here is the manner in which the doctor's report invokes the imagery of mobility to condemn her. She is not a seventeen-year-old girl. She is a 'raggarbrud', who, according to the report, seems to feel compelled to follow the 'the law of the raggare' rather than the (moral) law of society. She is in this sense understood to be completely 'out of place' (or to be lacking a moral place) to the extent that she stands not only outside of the home and traditional gender roles, but even outside of the law of society as a whole. The label 'raggarbrud', and its implicit illusion to mobility are enough to establish her presumed (excessive? improper? uncontrolled?) sexual activity – and perhaps more importantly, her lack of an ethical–moral understanding. Without these invocations of the (im)morality of mobility, the doctor's vague references to the fact that she 'probably' or alternatively 'obviously' has sexual relations, would seem flimsily thin at best. However, the unquestionable 'fact' of her immorality hardly seems to need excessive proof because it is derived as much from her association with mobility and other mobile youths as it is by her actual actions.

The consequences and outcome of this particular case are extreme. The threat of sterilization was not a fate hanging over the heads of the overwhelming majority of the women who associated with raggare. On the other hand, the moral condemnation explicit in this case was a reality which most of these women met and with which they had to contend. Working to explain why any woman would ever expose herself to this scrutiny and the hostility which followed it, other researchers have pointed out that in the life around the American car women could find an opening to a new modern identity which partially worked to distance these women from traditional gender ideals and notions concerning a woman's place in society (Bjurström 1990: 216; Turesson 1987). Nonetheless, they moved in a cultural sphere which was extremely male-dominated, and in which the vulnerability of their own position was tangible on several fronts.

At the same time, however, it should be pointed out that not everyone agreed that the writing of a *raggare* paragraph, and the mobilization of the police, social authorities, and mass media constituted the best way to deal with the situation, but as the debates about the new paragraph raged, *raggare* maintained their position as one of the most talked-about groups in society. And while the issue of sexuality was followed intensively, it would be inaccurate to explain the controversy around *raggare* in only these terms.

Returning for a moment to the men around the cars, the narratives of the mass media and middle class also played upon and enhanced other images, which contributed to a broader stereotypical narrative addressing not only issues of sexuality, but also intoxication and various forms of violence. Here, the fact that young men drank and then got into fights was in and of itself nothing new, but the narratives about *raggare* made them seem different. They formed 'gangs' with names reminiscent of American films, started rumbles at car races, and did not hesitate to turn on the police. For most Swedes, the whole thing smelled and felt too much like Marlon Brando in *The Wild One*. Indeed, *The Wild One*, and other Hollywood films such as *The Road Devils* and *Jailhouse Rock*, were sources of inspiration cited at times by *raggare* themselves (*Bildjournalen* 1959b: 19; Bjurström 1987: 41). But like Brando in *The Wild One*, popular narratives about *raggare* incorporated an element of unpredictability which made them particularly distressing in the minds of many men and women. With or without alcohol, *raggare* possessed the potential to simply come sweeping down on innocent and unexpecting bystanders:

> A brilliantly painted luxury car forces another vehicle on the Uppsala highway to the edge of the road. A few young men hop out of the luxury car and go to the other car. One of them jerks open the door of the hailed car. He says, 'Listen old man, we're gonna teach you to drive properly and leave enough room when we're going to pass. You don't own the right lane.' And then the other driver, a middle-aged man, is punched in the face. The young men climb back into their own car and disappear. On the back of the car there hangs a sign 'The Road Devils' (*Bildjournalen* 1959c).

Youths owning American cars were summarily labelled as *raggare* and they were portrayed as dirty working-class individuals who, if not directly pathological, at least chased young women, drank heavily, started fights, and seemed to often appear from nowhere only to vanish as quickly as they appeared (*Bildjournalen* 1959c: 9 and 1959d: 18; *Raggarungdom* 1962: 27 ff.)[12] Successively, the American car was transformed via these narratives into a signifier of potential danger and moral decline (LUF M20517: 3 and M20555: 2). The result was that as youths acquired the American car, the middle class increasingly came to criticize it.

However, for the youths themselves, the cars had a very special allure, the most important of which was undoubtedly the free space they provided. Writing on life in Stockholm in the 1950s Swedish author Per Anders Fogelström, has noted that most *raggare* were, contrary to most rumours:

> Pretty ordinary and quite decent youths who, in the life as a *raggare*, found some excitement and adventure in an otherwise rather boring daily life . . . Most of them probably lived in cramped housing conditions with their parents. The majority of them lived in the new suburbs which were not particularly fun; they needed to come into the city and drive around on *Kungsgatan* [a central street in Stockholm] for a while (Fogelström 1968: 223).

Figure 5.3 Despite popular stereotypes, it is not just teenage boys who own and admire older American cars. Photo: Tom O'Dell.

By escaping into the American car, Swedish youths escaped from claustrophobic living conditions and parental control (cf. Swärd 1993: 157). They obtained a free space in which they could express themselves and create their own identities. Once in the car the *raggare* was no longer the son of Mr and Mrs Svensson, but the guy in the Ford Crown Victoria: via the American car, the *raggare* gained distance from his family.

At the same time as the working-class youth obtained his own space and distance from his family, however, he seemed to be closing the socio-economic distance between the working and the middle classes. Previously, the automobile had been a status symbol reserved for the well-to-do (Bjurström 1990). However, this was clearly no longer the case, as working-class youths, with their newly acquired economic capacity, were attracted to the larger American car. The acquisition of the American car was further facilitated by the fact that the car tended to depreciate rapidly in economic value, making it easier for youths to purchase either a relatively new car, or even an older dilapidated model when economic realities forced concessions upon one's dreams.

Clearly, as was the case with jazz in the early 1940s (Frykman 1988), the threat of sexuality linked to the car caused much discussion and a moral panic. However, beyond this, the car drew attention because it so concretely symbolized the increased freedom of working-class youths and the ever-changing socio-economic order which this implied (cf. LUF M20546: 7–8). And here it can be argued that the union of youths and the American car epitomized not just a new age of consumption, but quite simply a new age of too much consumption and with it, too much mobility: cultural, social, and economic as well as physical.

In order to maintain its distance from the working class, the middle class was more or less forced to forfeit the American car as a status symbol. Simultaneously, this became easier for them to do as the American car developed incongruously with the predominant aesthetics of Swedish modernity; it increasingly came to represent the negative potential of Americanization in terms of violence, classless gaudiness, superficiality, and hedonistic consumerism. At the same time, *raggare* became infamous in the discourses of the mass media as the largest problem group among youths, and the most disturbing example of Americanization. Indeed, *raggare* so completely appropriated the American car from the middle class that still to this day, no banker or high school teacher would ever consider owning such a car.[13]

Linkages and Transformations in the Transnational

In this way the American car remains rather unique as a symbol. While members of the middle class may find the idea of owning an old Chevrolet untenable, most people continue to be extremely curious about these automobiles and their owners. In line with this, it is not uncommon to hear appreciative comments from pedestrians as an American car glides by on a warm July evening. Similarly, the national newspapers still regularly cover the largest American car meet held every summer outside of Stockholm, while local newspapers report on the smaller meets held throughout the country. Together, they monitor these meets half expecting to find the violence, sexuality, and hedonism of yore. However, year after year, these newspapers can only report, with a sense of surprise that, 'this year there was not much commotion as a result of the fair. It's been quite a while since things went this smoothly and quietly' (*Jönköpings-Posten* 1992: 2). Despite over forty years of activity, *raggare* and their cars are, in short, still news worthy. In comparison, it can be noted that while streamlined refrigerators and vacuum cleaners may have captured the imagination of many Swedes in the 1950s, they are now all but forgotten. The American car remains vital as a symbol and this works as a potent reminder that the automobile is much more than just another mechanical, mass-produced commodity among others. It is different in some sense.

In opening I directly indicated the significance the domestic sphere has as an arena of self-expression and identity in the Swedish context. Still to this day, the home in Sweden has a central position in everyday life, not only as a private place in which familial matters are worked out, but also as one of the principle spheres in which friends and neighbors socialize.[14] In Sweden, bars, pubs, and restaurants do not play the same social role as they do in other Anglo-American contexts, although cafés do work more as a public arena in which people can meet and socialize, especially in larger cities and towns. Nonetheless, the public sphere continues to be a slightly more formal, impersonal, and subdued arena of cultural interaction in Sweden. In this context, I would argue that the symbolic strength of the car is accentuated by its ability to bridge, and perhaps even subvert, the distinction between the private and public spheres, at least as it is related to young men and women.

More than just an object, the car is a room in and around which working-class youths can develop their own modern identities. Dressed in blinking Christmas trees, American flags, and foxtails, the cars contest the dominant aesthetics of 'good taste', and work as personal statements through which class tensions can be negotiated. However, as I have argued above, in purchasing the car youths have also bought into a series of dreams, ideals, and romanticized fantasies linked to the car. For many, the perceived freedom of hitting the road – American style – is an important aspect of the entire symbolic package associated with the car (cf. Eyerman and Löfgren 1995).

Clearly, since the 1950s, the car has been a means of escaping parental supervision, but as my research among *raggare* today indicates (O'Dell 1997), 'hitting the road' brings with it a series of expectations as well as a tingling sense of anticipation which derives from the feeling that almost 'anything can happen'. For these youths, the car constitutes an arena around which it is possible to test and slightly stretch the limits of 'acceptable behaviour' as they drink, sing, scream, play loud music, and shout greetings to unknown pedestrians who are themselves out for an evening stroll. Here, a fun evening usually does not include more than driving around the town square for several hours, and stopping for a hotdog. However, in this movement exists the potential of meeting

Figure 5.4 After a long night of festivities, a young man seeks out a cool and shady spot to take a nap in the middle of the 1995 Power Big Meet. Photo: Tom O'Dell.

other friends, and making new acquaintances of both sexes. In this sense, it is possible to liken the car to a mobile family room or kitchen – a semi-public sphere in which friends congregate and socialize.

For *raggare*, the mobility provided by the car was (and is) an inseparable and central component in the cultural construction of their experience of adventure. Through mobility it was always possible to shift from public to private spheres by moving from the town centre with its crowded streets to the sparsely populated countryside with its secluded forest roads. This was exactly the issue which caught the attention of authorities, parents, and the mass media, who did not hesitate to liken the car to a rolling bedroom, or brothel. From this perspective the car was an extremely private, sensuous, and confined room, and obviously, the sexuality of the car and the privacy it could provide a young couple worked as much as anything else to tantalize the imagination of the young men and women who drove around in these machines.

However, from society's perspective, mobility was a morally laden problem. Accordingly, *raggare* were often portrayed as roving predators who could appear from nowhere, cause havoc, and disappear again. Similarly, the women in the cars were easily stigmatized as promiscuous, or in some way indecent. These young men and women seemed to exemplify a new rootless generation living by their own rules, and lacking a well-anchored and stable moral constitution. The car worked to accentuate these generation-bound concerns, but they were further exasperated by the transnational flow of images of rebellion coming from the United States and Hollywood. The car provided youths with the possibility of escaping society's panoptic eye for a moment, but Marlon Brando and James Dean contributed, as much as anything else, to the sense that the car was an arena of imminent danger and moral degradation. At the same time, however, it should be emphasized that the product of these working-class youths' activities is today one which (to the surprise of most Swedes) has no absolute equivalent in the United States. The manner in which these youths use the cars and manipulate the symbols around them is entirely Swedish (O'Dell 1992, 1992 and 1997). Their interest is not American culture per se, but the agitation of the middle class (although they would not explain their actions in quite these terms), and to this extent it can be said that their own aesthetic values are defined in contrast to those of the middle class.

In the end, it is the combination of these elements that makes the American car different, not only from most other mass-produced commodities, but also from all other cars. In the Swedish context, a 1959 Volvo PV 544, for example, does not provoke the same symbolic associations as a 1959 Chevrolet Impala, and never will. The Volvo is romantically and nostalgically seen as 'the people's car' and associated with the Social Democratic project of the people's home. It symbolizes history, cultural heritage, and Swedish roots. The Impala is still an intruder and symbol of youth rebellion, Americanization, and the socio-economic mobility of the working class. Through its public consumption the American car continually re-invokes a history of class tensions which distinguish it from so many other commodities in society. These tensions are at times consciously manipulated by *raggare*, but they are also projected on to the cars by bystanders who may be able to admire the cars at a distance, but who still warily watch these finned foreigners glide by, awaiting the first sign of potential danger or violence.

References

Appadurai, A. (1986), 'Introduction: Commodities and the Politics of Value'. In A. Appadurai (ed.), *The Social Life of Things: Commodities in Cultural Perspective*. London: Cambridge University Press, pp. 3–63.

Arnstberg, K.O. (1989), *Svenskhet: Den kulturförnekande kulturen*. Stockholm: Carlssons.

Bildjournalen (1959a), 'Titta . . . Där Har Vi Ragg', 1: 2–3.

—— (1959b), 'Djävlarnas Två Ansikten', 30: 19.

—— (1959c), 'Men Hur Tuff Är Sven i Verkligheten?' 26: 9.

—— (1959d), '"Den Där Djävla Snuten" Vill Säga Er en Sak', 32: 18.

Bjurström, E. (1987), 'Bilen och Motorcykeln i Ungdomskulturen'. In *Ungdom och Traffik: En Omöjlig Kombination?* Stockholm: National Förening för Traffiksäkerhetens Främjande, pp. 33–53.

—— (1990), 'Raggare: En Tolkning av en Stils Uppkomst och Utveckling'. In P. Dahlén and M. Rönnberg (eds), *Spelrum: Om Lek, Stil, och Flyt i Ungdomskulturen*. Uppsala: Filmförlaget, pp. 207–228.

Broberg, G. and Tydén, M. (1991), *Oönskade i Folkhemmet: Rashygien och Sterilisering i Sverige*. Stockholm: Gidlunds Bokförlag.

Clifford, J. (1997), *Routes: Travel and Translation in the Late Twentieth Century*. Cambridge, Mass.: Harvard University Press.

Daun, Å. (1996), *Swedish Mentality*. University Park: Penn State University Press.

Eriksson, R., Andreason, O., Aldman, B., Christer, V., Ullén, J. (1959), 'De Tre Stora: Chevrolet, Ford, Plymouth', Teknikens Värld, 16: 24–33.

Eyerman, R. and Löfgren, O. (1995), 'Romancing the Road: Road Movies and Images of Mobility'. *Theory, Culture & Society*, 12: 53–79.

Fogelström, P.A. (1968), *Stad i Världen: 1945–1968*. Stockholm: Bonniers Pocket.

Friman, H., Henschen, H., Högberg, L., Silvén-Garnet, E. and Söderlind, I. (1991), *Storstads Ungdom i Fyra Generationer*. Stockholm: Tidens Förlag.

Frykman, J. (1981) 'Pure and Rational. The Hygienic Vision: A Study of Cultural Transformation in the 1930s. The New Man'. *Ethnologia Scandinavica*, 36–63.

—— (1988), *Dansbaneeländet*. Stockholm: Natur och Kultur.

—— (1989), 'Social Mobility and National Character'. *Ethnologia Europaea*, XIX: 33–46.

Frykman, J. and Löfgren O. (1985), *Modärna Tider*. Malmö: Liber.

GM Revyn (1958), 'Den Svenska Kvinnan Har Smak', Fall: 2–3.

—— (1960), *'Tittskåpen Blev Folkhem: Möte med Lena Larsson'*, Spring: 12–17.

Gullestad, M. (1991), 'The Transformation of the Norwegian Notion of Everyday Life'. *American Ethnologist*, 18: 480–499.

Hine, T. (1986), *Populuxe*. New York: Alfred A. Knopf.

Jönköpings-Posten (1992), 'Dåligt Väder Gladde i alla fall Polisen'. April 21: 2.

Löfgren, O. (1988), 'En Svensk Kulturrevolution? Kampen mot Soffgruppen, Slipsen och Skåltalet'. In O. Löfgren (ed.), *Hej, Det Är Från Försäkringskassan*. Stockholm: Natur och Kultur, pp. 174–206.

—— (1990), 'Consuming Interests', *Culture and History*, 7: 7–36.

—— (1991), 'Att Nationalisera Moderniteten'. In A. Linde-Laursen and J. O. Nilsson (eds), *Nationella Identiteter i Norden*. Eskilstuna, Sweden: Nord, pp. 101–48.

—— (1993), 'Swedish Modern: Nationalizing Consumption and Aesthetics in the Welfare State'. Working paper presented at the workshop, 'Mass Consumption, Civil Society, and Changing Political Systems', Rutgers Center for Historical Analysis, April 16–17.

Marcus, G. (1998), *Ethnography Through Thick and Thin*. Princeton: Princeton University Press.

O'Dell, T. (1992), 'Myten om "amerikanaren" och Sverige', *Kulturella Perspektiv*, 3/4: 32–39.

—— (1993), '"Chevrolet . . . Yeah, That's a Real *Raggarbil*." The American Car and the Production of Swedish Identities'. *Journal of Folklore Research*, 30: 61–74.

—— (1997), *Culture Unbound: Americanization and Everyday Life in Sweden*. Lund: Nordic Academic Press.

—— (1998), 'Junctures of Swedishness: Reconsidering Representations of the National'. *Ethnologia Scandinavica*, 28: 20–37.

—— (1999a), *Nonstop! Turist i upplevelseindustrialismen*. Lund: Historiska Media.

—— (1999b), 'Metodens praktik: uteslutningens poetik'. In T. Damsholt and F. Nilsson (eds), *Ta fan i båten: Etnologins politiska utmaningar*. Lund: Studentlitteratur, pp. 58–79.

Raggarungdom (1962), '*Utredning angående de s.k. raggarproblemen i Stockholm. Verkställd av en av Stor-Stockholms planeringsnämnd tillsatt kommitté. Stadskollegiets utlåtanden och memorial*.

Rojek, C. and Urry, J. (1997), 'Transformations of Travel and Theory'. In C. Rojek and J. Urry (eds), *Touring Cultures: Transformation of Travel and Theory*. London: Routledge, pp. 1–22.

Rosengren A. (1997), 'Raggare och Jänkebilar'. In B. Bursell and A. Rosengren (eds), *Drömmen om Bilen*. Stockholm: Nordiska Museet, pp. 148–167.

Schoug, F. (1991), 'Välståndets Missionärer'. *Nord Nytt*, 44: 54–66.

Se, (1961), '30 Minuter som Skakade Lyssnarna'. 18: 40, 42, 44.

Spångberg, A. (1962), *Raggare*. Stockholm: Wahlström & Widstrand.

Swärd, H. (1993), 'Raggarflickorna'. In H. Swärd, *Mångenstädes Svårt Vanartad . . .: Om Problem med det Uppväxande Släktet*. Floda: Zenon, pp.151–215.

Turesson, M. (1987), *Raggarbrudar*. Unpublished stencil, Uppsala: Department of Ethnology, University of Uppsala.

Virilio, P. (2000), *Polar Inertia*. London: Sage.

Wolff, J. (1995), *Resident Alien: Feminist Cultural Criticism*. Cambridge: Polity Press.

Other Source Material

LUF: Responses to Questionnaire #181 "The Car", Folklife Archives, University of Lund.

Notes

1. In addition, it should be noted that one of the underlying intentions of this chapter is to follow the American car through a series of different cultural

settings in an attempt to better elucidate the manner in which it is constantly reworked, producing different cultural consequences from setting to setting. Here my thinking is inspired by Appadurai's work on the social life of things (1986) as well as Marcus' discussion of the need to better understand cultural processes in terms of multi-locality (1998).

2. I myself have translated into English this and all other material quoted in this chapter from the original Swedish.

3. See for example Arnstberg 1989; Daun 1996; Frykman 1989; Löfgren 1993.

4. The statistical information included here, and further below, has been obtained from the Association of Swedish Automobile Manufacturers and Wholesalers. In these sales statistics I have only included those for new cars which were actually imported to Sweden from the US. The reason for this is that Ford in particular presents a problem since it did produce and import cars to Sweden from other European countries such as England, France, and Germany. However, European-produced Fords such as the Taunus and Anglia tended to look more like other smaller European cars, and as my interview material indicates, Swedes tended to perceive these cars as just that: European-produced Fords, which were in some way different than American-produced cars. They are therefore not included in my statistics over cars produced in the US and sold in Sweden, even though one could argue that they were products of an American automobile manufacturer.

5. My discussion of the middle class is a bit ambiguous here. It is ambiguous due to the difficulty in clearly defining the middle class. To generalize slightly, it seems as though the middle class is much more of an economic construction in the United States than it is in Sweden where it is defined more in terms of education and occupation. For my purposes here, I think it is sufficient to say that in America there is a large majority of people who are neither extremely wealthy nor extremely poor (in their own eyes); they define themselves in terms of, and in contrast to, one another: this is the American middle class. I see the Swedish middle class as primarily but not exclusively consisting of university-educated individuals, and civil servants. These are people who are not manual labourers and do not identify themselves as what in Sweden is commonly referred to as 'ordinary workers'. Likewise, I do not consider people who occupy highly influential managerial positions in financial and other industries as members of the Swedish middle class.

6. *GM Revyn* is a magazine published by General Motors Nordiska AB to promote the sales of GM cars in Scandinavia.

7. A car, implicitly understood to be American, driven by *raggare*.

8. *Bildjournalen* was a popular magazine, generally full of large photographs, directed at teenagers, and often featuring a wide array of articles about various film and pop stars.

9. 'Raggarbrud' is a term used to refer to a woman who associates with *raggare*. Formally, 'brud' is the Swedish word for 'bride'. However, its slang usage lies closest to the English word 'chick' or 'bird', and refers in a generally non-flattering manner to a woman. In line with this usage, the word 'raggarbrud' is stigmatizing and derogatory.

10. For reasons of space I have edited this passage in order to shorten it. Most of the material I have omitted is a repetition of the points demonstrated in the material quoted in reference to the young woman's intelligence and morality. All of the omissions indicated in the text are my own.

11. The first law concerning forced sterilization was passed in 1934 and put into practice as of 1 January, 1935. From the beginning, the decision to sterilize a person – with or without his or her consent – could be made via an appeal to one of two factors or preconditions. Here, reference was made to a 'social indicator' and a 'eugenic indicator'. 'The preconditions were that the person who was to be sterilized was, for all intent, not capable in the future of being responsible for the care and well-being of a child (the social indicator), or due to genetic disposition would transmit mental illness or mental deficiency (the eugenic indicator)' (Broberg and Tydén 1991: 72). In 1941 the law was broadened, allowing for the forced sterilization of people who were thought to maintain an asocial lifestyle or way of living, as well as for people who for medical reasons (often with an appeal to perceived risks to the individual's own health) should not have children. Of all the people sterilized 93 per cent were women (Broberg and Tydén 1991: 93ff.).

12. Here it is interesting to note that *raggare* and the women associated with them were by definition other than normal. In *Raggarungdom* for example, the youths belonging to automobile clubs are compared statistically to another category of peers called 'Normal group 1958 boys'. As mentioned earlier, *raggare* had their supporters, people who argued that these were just youths having a good time, but the tone of the general discourse around them clearly marked them as a problem, and defined them as a morally questionable group in need of help.

13. European-built Fords and Opels are an exception to this rule, and are not in any way considered to be *raggarbilar*. For a more detailed explanation of the appropriation of the American car by Swedish youths see O'Dell: 1997 and 1999b: 68ff.

14. A strong parallel can be drawn here to the Norwegian context in which Marianne Gullestad explicitly asserts the central significance which the home plays for Norwegians as a point of departure for the activities of everyday life and underlines this point in the following manner: 'What I want to emphasize is the centrality of the home in Norwegian culture. There is a truly . . . sacred sense to how Norwegians regard their homes . . . The notion of closeness is

thus a spatial metaphor for specific relational qualities. For Norwegians the home is a concrete material manifestation of special relationships. Working on the home both creates and symbolizes closeness. In the idea of a good home, a variety of ideas about place, people, and relationships are brought together' (1991: 490–1). In this sense, we find in Norway as well as in Sweden a clear distinction between the home, intimacy, and identity on the one hand, and the more formal and impersonal arena of public life, in which individual distinctiveness is toned down.

Driving, Drinking and Daring in Norway

Pauline Garvey

Introduction: Kari and Eva

The general image of Norwegian society both internally and to the outside world is dominated by influences such as rationalism (Witoszek 1997) and citizenship. It is thought of as generally law-abiding, egalitarian (Graubard 1986; Gullestad 1992), with a strong affection for the countryside (Nederlid 1991) and lack of affectation. This paper aims to explore practices of transgression or abandon within the normative and conventional settings of young working-women's lives and their day-to-day routines. The fieldwork on which this paper is based was carried out in Skien, a town with approximately 49,000 inhabitants, which is located in south-east Norway. Ethnographic research was carried out during 1997/98 and focused upon women in their twenties, who saw themselves as both ordinary and typical, although perhaps with a little lower income and slightly less in the way of prospects than would be typical of an affluent country such as Norway.

In the following I explore drinking and driving as separate events and as transgressive media among my informants. As I draw extensively on case studies, it is important to note that the material is ethnographic and not just anecdotal. That is to say, the behaviour I describe belongs to a recognized genre within the milieu where I carried out fieldwork and therefore represents at least a section of the Norwegian working class. From an analysis of case studies, I look at how, and in what situations the car and alcohol possess similar qualities in providing media for transgressive behaviour and why national trends, social

133

attitudes and state regulation has resulted in imbuing these media with this potential in certain and contained contexts.

Kari works in a kindergarten. She is 25 years old and lives alone, but being an extrovert she often suffers from ennui or loneliness when at home. When I first met her, her childhood friend Hilde had just settled in with her boyfriend and was busy with all that that entailed. One morning she explained to me how they had a long tradition of getting out together and just then both felt they could do with a break: 'Are you sure you want to come' I was asked as we set off, 'we can be a bit crazy'. For such excursions the car is critical. But this may not just be a case of driving around. On this occasion, the day started slowly for Kari, and after catching up on the week's episodes of 'The Bold and the Beautiful', she lolled around on her sofa, gradually finding the domestic atmosphere increasingly oppressive. We collected Hilde in Kari's battered Ford and proceeded to the outlying streets which encircle the town. Being a quiet Sunday afternoon it took some time to make an impact on a few lone pedestrians or motorists. Distances need to be gauged quite precisely, close enough to the town to make an impact, but far enough to avoid 'getting a ticket'. They started quietly at first, initially just pretending to stall the car and then laughing hysterically when other drivers hesitantly tried to overtake. Hilde knew the routine and opened her window whilst shouting ohhhh, aaahhh or other such yells at shocked pedestrians as we passed by. The excitement gradually mounted and with it Kari and Hilde grew more daring. Kari began driving on the wrong side of the road, each complimenting the other on being 'crazy' while swerving suddenly in the direction of oncoming cars. 'This is Norwegian humour' she quipped to me in the back seat, suddenly changing direction at a T-junction and near-missing a Volvo.

Such behaviour is not unprecedented for Kari and she can describe a number of similar incidences, with different friends. On another occasion for example, she was cruising town with Gro:

> . . . and we held up traffic in the middle of town. I laughed so much, I thought I was going to die. I got out and opened the (engine) and pretended I knew what I was doing and as soon as someone came to help I shouted to Gro 'try it now' and then it started of course because there was nothing wrong with it. I laughed so much, I have not had that much fun in a long time.

In conversation with Kari about similar events, I found her comments underpinning the use of the car as opening the sluice gates which inhibit her expression of moods and feelings. At times nothing else would really do. She describes a conversation about this with her friend Eva:

> That is the best thing I know when I am angry or frustrated, to take the car. She (Eva) said, 'well I can't as it is too dangerous. Like if you are really pissed off and you think OK, life sucks, it really stinks and you are in your car going 60km an hour and just take one hard turn with your steering wheel into something then . . .'
>
>   You are doing it deliberately but you don't mean it, but you are so angry you can't think clearly so you just think, OK let's try it, God damn you. So it can be dangerous, but that is what is good about me because when I am driving I block everything out. I think I know a lot – or I remember talking to others, who say when they need a break they take the car for a ride, it just helps I don't know why.

In an analysis of car advertisements in Sweden, Hagman (1994) illustrates that key images drawn upon relate to effectiveness, predictability and independence. One particularly successful campaign, for example, ostensibly placed the values of 'rationality' in contradistinction to those of prestige. This campaign, originating in the 1960s and one of the most successful of its time, pronounced that the Renault 4L would appeal to the discerning consumer for whom a rational (rather than a prestigious) choice would be paramount. Hagman (ibid.) argues that by thus denigrating a form of showy prestige, it functioned to associate the Renault with a more culturally benign and less explicit form of kudos. Tom O'Dell in chapter Five of this volume, on the other hand, provides an example of how historical trajectories reinforced associations between the car and transgressive behaviour in Sweden. In his case this depends upon discriminating a particular class or car equivalent to the *råne* in this chapter, and therefore while the transgressive use of the car might not appear so unexpected in certain contexts, in countries such as Norway where order and safety are constantly stressed, Kari's and her friends' behaviour comes across as particularly unexpected.

Kari and Hilde's excursions seem to suggest a 'flip' from normative and acceptable action to something that is clearly constructed in marked opposition. In order to understand the use of the car in creating these transgressive moments, it should be considered within the wider social setting – which is, the focus of ethnographic enquiry. Within this larger study there is one other activity that seems to have many of the same qualities, the same 'flip' into transgressive modes that are pitched against acceptable and expected norms. This contravention of social codes lies not in alcohol per se, but rather in the particular form and nature of drunkenness as it is performed and understood in this cultural context. To exemplify this we turn from Kari to Eva.

Eva is a 25-year-old woman. Her flat occupies the upper part of a privately owned house and consists of one bedroom, a kitchen, a bathroom and a living room, which she keeps immaculately clean. Eva is fanatical about harmony and symmetry; everything has to be in its place, even to the point that if she eats chocolates, they have to be eaten in rows because she hates to see gaps all higgledy-piggledy: 'even if all the others are all there, if there is a space in the middle, I keep looking at the gap and it gets on my nerves.' She is endlessly sorting and rearranging her furniture and decorative items keeping them 'harmonic'. 'If the (stuffed) mouse was over on one side and I saw that, I couldn't leave it like that, I couldn't go to sleep before I had stood it up again.' She admits that this in part relates to a sense of insecurity and lack of confidence in social relations. Control within her home assuages insecurities elsewhere.

Despite the extremes to which Eva imposes order on her home, she relaxes these rules when she hosts drinking parties (*vorspiels*) of up to fifteen people, mostly in their 20s and 30s. These evenings follow a regular pattern. Typically the evening starts quietly but soon becomes lively as the drinking and dancing gets under way. The atmosphere is playful; at certain points war is declared over music choices, the dancing becomes buoyant and may move onto tables. Although usually reticent, Eva acts the hostess in command throughout. She dances, urges the others on and courts attention by bizarre behaviour such as spilling drinks. Although she is known for her strict orderliness, at the *vorspiel* bottle caps and later larger items can be thrown around the room with impunity and even out the window.

The starting point of this chapter is the very exceptional nature of both driving and drinking within such exercises of transgressive daring. They make up two dominant media for such transgression. In both cases they seem to represent a dramatic 'flip' across from accepted and conventional behaviour, and yet both are carried out in social settings and clearly appear to conform to 'normative' genres of social con-travention. They have a particular quality that makes them stand out in a setting framed as rational (Witoszek 1997) and practical (Gullestad 1989), such that these actions appear as the two bright colours on an otherwise grey landscape. Ideas of cultural daring are framed in particular and nuanced ways and although alcohol or cars are universal vehicles for transgression, the ways through which this is understood or framed is highly textured. For example Gefou-Madianou's *Alcohol, Gender and Culture* (1992) illustrates the various ways alcohol is used as a social tool and, for example, that Irish fishermen exhibit a 'virtual cultural requirement to drink' but also to maintain control (Peace 1992: 171): 'fishermen imbibe heavily and become somewhat inebriated . . . they do not thereby lose control over their immediate circumstances or indeed abandon their sense of judgement.' (Ibid.: 179).

In Norway the transition from sobriety to drunkenness is particularly framed through the institution of the *vorspiel*. Unlike the example of the Irish fishermen, this process seems much more a form of sudden movement from one state to its opposite, and confirms the socially embedded nature of drunkenness, which in a Norwegian setting could be argued to be premised on values such as abandon (Sande 1996; Sørhaug 1996), or performance and display. In other words it is not just a case of drunkenness per se, but specifically Norwegian drunkenness. While in other societies the relationship of sober to drunk may be one of gradation, in Norway it is experienced as two opposed states. So in the same way that being drunk has a particularly Norwegian inflection, what Kari and her friends demonstrated is not just car transgression but a specific Norwegian genre of car transgression.

The argument of this chapter is that the reason for which driving and drinking have come to occupy a similar niche within a sector of Norwegian society only makes sense when situated within the history of these two forms. The conclusion will be that to understand the contemporary role of the car we have to take note of the way the Norwegian state responded to the development and spread of the car itself. In short the Norwegian state treated the car as though it were a kind of 'alcohol' and this may well explain why some Norwegian people have constructed an oppositional role for the car in which the car became used as though it were a kind of 'alcohol'.

Domesticity and the Transgressive Potential of the Car

With a traditional automobile industry in Norway lacking, the history of the automobile is one of consumption rather than production; of post-war state measures, import and sale restrictions and state attempts to confine social life to 'necessities'. Moving steadily from being highly restricted, declaimed as a luxury and associated with 'the old class society' (Østby 1995), the car has become an essential personal item which has grown with the increasing wealth and disposable income of the Norwegian populace. The car therefore has been, in many ways, the harbinger of modern Norway. Through its integration in society, it has been both welcomed as a symbol of affluence and accused of being a threat to traditional social structures. Despite early restrictive state policies, popular demands necessitated the car's integration and acceptance, although given its association with modernism, affluence and consumerism, the optimism which greeted its arrival has soured somewhat in recent years. It is not surprising then that public concern relating to wealth or conspicuous consumption has come to focus on cars and car culture in general. Replete with moral purpose, a national newspaper heading warns that 'young people have stopped worshipping Jesus, and started worshipping the BMW' (*VG*, 15/03/1997). The car is attacked as a status symbol within an egalitarian society and as materialism within a Puritanical society. In this paper, however, the emphasis will be less upon these concerns, since the ethnography was carried out among a less affluent segment of the population. Within a context of young women settling into family life, the car is ubiquitous within a domestic setting, carrying the routines and responsibilities of domesticity.

Domesticity still forms a major category in the formation of individual and collective identity for my informants, who consisted of single women living alone or mothers of young families. Extending far beyond the confines of the family home, domesticity encompasses gender roles, parental and child relations, attitudes to socializing and romantic relationships. Almost all aspects of work and social life are posited in relation to domestic ideals or fall-offs from these ideals. On the one hand, definitions of home incorporate ideals of security and love. The way people experience domesticity as a way of life, however, incorporates a reiteration of routine and chores, familial cares and the banal difficulties involved in negotiating multiple relationships as a family. The car functions as both a primary tool in the negotiation of domestic chores, and as a medium of escape from those same chores. Most immediately, one associates the car's role as a form of 'escape' with teenagers, whose first car often signifies a movement to adulthood and an exploration of their own identity (see Löfgren 1994: 52–3). Speaking more generally, while most instrumental forms of driving are routinized, the car dominates leisure where '. . . car-ownership has come to mean freedom in terms of greatly increased mobility and independence of rails and other public means of transportation which demand planning and structuration of leisure and holidays' (Sørenson and Sørgaard 1994: 21). This ideal of the holiday as escape applies to most households, and not just the young.

While rationality or predictability are commonly associated in car advertisements, it is perhaps unsurprising that this set of values is inverted and challenged through the same form of material culture. This is particularly so in view of the underlying associations of domesticity which the car embodies. We see this below in the description of humour and transgressive behaviour where I explore the inversion of key symbols of domesticity by manipulating the significances which they usually inhere. The car for example is a vital tool in the running of banal family errands which, in certain circumstances, is transmogrified in order to become a tool of humour, anonymity, and abandon. This is strikingly the case in examples whereby the car is appropriated by subcultures, youth groups or working-class males, as illustrated effectively by O'Dell (1997 and this volume; also Lamvik 1996; Rosengren 1994), but can also be employed by average home-owners in their play on domestic values to realize other forms of behaviour.

In domestic routines the car is a gendered artefact; women often take responsibility for the upkeep of the home, while men shoulder the responsibility of the family car. Although this division is not an absolute, it does throw light on the specific importance of the car for Kari and Hilde for whom the car comes to represent the 'other' to the home that Hilde had been busy establishing. The physical distance from the home that the car allows, coupled with the reduced likelihood of being recognized, enhances their sense of abandon. We see this in the following example of a middle-aged woman. Karoline who would on no account condone Hilde and Kari's behaviour, on her own trips away likes to participate in a particular form of humour which is spontaneous and impulsive. Karoline, a highly responsible 53-year-old nurse, also associates the car with transgression, in that it takes her away on weekend trips with her work colleagues where they sing songs, play games and end up 'howling with laughter'. The car here is important as much because it is the technology that creates distance from the home in physical and in symbolic terms. It provides a space where people caught within routines of home and domesticity realize the kind of fantasies that, in their book *Escape Attempts*, Cohen and Taylor (1992) argue are central to everyday life. The reckless gales of laughter signify that the proper abandoned state has been realized. Despite the routinized quality of these trips and the conventional forms of transgression which they engender, an element of the unexpected or spontaneous imparts an atmosphere of freedom for the participants.

The transgressive use of the car builds upon a more gentle sense of difference and distance that the car permits. For example, the car may be used as a bulwark against the threat of social isolation, which is also a feature of domestic life. This is evident in the example of Dorothy, a 24-year-old young woman, who is unemployed and lives alone. Due to her rather solitary existence she is heavily dependent on her car and professes to find something to do in town every day, just to give her the excuse to get out and drive somewhere. Other than these errands which have become in themselves routinized, the car fulfils a form of vicarious sociality and she especially likes to go driving when she is tired of being home and unable to find company. She describes how she saw an advertisement for a car that had a computer installed in which it would tell you if there was something wrong with it. She jokes that this would be funny as she could talk to her car – and programme it to ask her if she had a nice day. The car then occasionally substitutes for absent social relationships and she drives when she is lonely or bored – frequently at night while listening to music. Although driving is a solitary experience for Dorothy, as she traverses unknown country roads and sings along to the radio, the excursions act as a pressure valve to release the oppressive isolation of long periods inside the home. When the quiet atmosphere of home becomes stifling, she has a means to escape and return when it is again tolerable or even welcome. The car therefore becomes an extension and an expression of her sense of agency which allows her to alleviate associated difficulties of living alone.

With Kari this configuration of the car in opposition to the home is quite explicit. As she puts it:

> Because if I am just sitting here, listening to music, I am thinking and I am fantasizing and thinking oh if it was like this or that, I could do this and that but if you are driving you have to concentrate on the road, so you hear the music and you can sing to it but you can't think like that as you have to concentrate on driving so it clears your head . . . A problem seems worse because you are so used to sitting here, I am always here and usually I am here on my own, I don't mind that but when you are on your own it is easy to feel sorry for yourself and think 'OK, I am here on my own again', but when you take your car you can always pretend to go somewhere the car is taking you from A to B and then back to A again but when you sit at home you are just staying at A, so just to get out.

From this point one can see the more extreme opposition that leads to her transgressive use of the car as an expression of anger or frustration that was illustrated in quotations earlier in this chapter.

Men and Modernity

If for women the key relationship in understanding the transgressive potential of the car is that with domesticity, then for men the equivalent would be the link to modernity. Sørensen and Sørgaard (1994: 4) point out that in Giddens's discussion of high modernity, technology remains unexplored. The clock, electronic communication and the car are mentioned but not analysed. These contribute to the dynamic nature of modernity by contributing to three modern qualities:

a. *Separation of time and space*: the condition for the articulation of social relations across wide spans of time-space, up to and including global systems.
b. *Disembedding mechanisms*: consist of symbolic tokens and expert systems. Disembedding mechanisms separate interaction from the particularities of locales.
c. *Institutional reflexivity*: the regularised use of knowledge about circumstances of social life as a constitutive element in its organisation and transformation (Giddens 1991: 20).

The car in Norway fulfils many of these criteria as outlined by Giddens. Essentially, modernity is defined as a post-traditional order. The car is strategically involved in the reorganization of time and space, in redefining the 'local' but also incorporating a universalizing awareness; including global environmental concerns. Equally, the car is an *expert system* in having a technological validity independent of the practitioners who make use of it (1991: 18). With the car, patterns of interaction and sociability take on new perspectives and, as we will see below, residential trends aligned themselves in accordance with the greater accessibility which the car facilitated. The car has also been involved in the third element of reflexivity, partly as evident in the radical changes in the Norwegian state's approach to it during the 1950s/60s/70s which will be discussed below, and more recently in the increasing environmental apprehension which surrounds the car today.

If the car is a modern symbol, the question remains as to which aspects of this symbolism are accentuated in the consumption and development of a specific car culture. Two which seem particularly relevant to male usage are speed and display. Speed is central, where configurations of freedom are often framed as intrinsic to the driving experience. Speed involves both defiance and control: defiance in transgressing norms and safety regulations, but control in commanding authority over one's fate. Therefore it seems unsurprising to read newspaper articles suggesting that it is neither the search of dangerous experiences or pursuit of risk which makes young drivers a danger to themselves and others, but rather an exercise in control which exacerbates the urge to speed (*Aftenposten* 22/08/97). The wish to master is crucial here and it is argued that 'a moderate amount of risk' can be interpreted as belonging to 'healthy, normal development' (*en moderat mengde med risiko hører med som en ingrediens i en sunn, normal utvikling*).[1]

These findings are supported by my informants, some of whom argued their right to speed in terms of their control over the car and their ability to avoid collision. In spite of his involvement in three road accidents, Erik, aged 19 years, maintained that his experience of speeding implied that he was a safer driver than others who might be unable to control their car at high speeds. Emphasis here was put on control and being able to maintain control in adverse conditions, such as speeding on country roads. The issue of youth speeding has become such a matter of public and official concern that various unconventional measures are suggested to combat this problem. A number of politicians in northern Norway even suggested offering an award of 10,000 NOK[2] for those who avoided traffic accidents (*Aftenposten* 20/11/97).

Erik's interest in cars is not, however, limited to speed. Like many young men he is equally interested in its potential for display. Erik drives a lime-green Toyota. He didn't actually buy the car but acquired it when he and a friend swapped vehicles. His job in a local fish and chip shop gave Erik few resources to invest heavily in his new acquisition; it had no windscreen or lights and initially made very slow progress. For months, one could hear him and his friends pushing the Toyota around the local residential streets, trying to get it started. After much expense and effort however, the car was fitted with a powerful engine and adequate windscreen and cruised the town on Saturday nights. Cars of this nature have been described to me as typically *råne*, a label which refers specifically to the elaboration of the car with additional lights, large wheels, powerful engines and details down to the *wunderbaum* trees dangling from rear-view mirrors. The explicit embellishment of the car and the attention-seeking antics of råne car owners is a common point of local amusement and Skien newspapers parody the driving style of the råne – with one arm resting at a 180 degree angle while the other grips the steering wheel and drums lightly to the deep bass throbbing from the car (*Varden* 23/01/98). Having a car of this calibre is of little use unless it is paraded and one finds cars circling the town centre on weekend nights (cf. *Varden* 23/01/98 for description of the råne trend in Skien). For some, the råne represents a sense of the vulgar and uncouth which is generated by the conspicuous show of flashy cars.[3] From the point of view of Erik however, the urge to revitalize old cars with high-powered equipment might relate to a young man's attempt to generate an image of himself as the owner of a more glamorous model. While the cars he emulates remain outside his price range, the qualities of 'toughness', power and fashion are projected on to the adjustments made to his own car and the context within which it is employed, such as cruising around the town.

The situation with young men parallels that of women. Here there is a general expression of aspects of modernity, such as display and materialism, but also mobility. These can become the foundation for more explicitly transgressive behaviour based around speed and courting danger. While gender brings out and clarifies certain traits it should be not exaggerated. Women also get 'high' on speed and young men are also concerned to establish their distance from domesticity and family more generally. Youth, freedom, danger, modernity are the immediate associations and attractions of the car, which in turn make it a facilitator of non-domestic values. This may emerge equally through the vulgarity of Erik's car as in the laughter of women trying to experience something that negates their daily life and routines. Superficially this can sound much like the US trajectory in films from James Dean to Thelma and Louise, but the situation in Norway could hardly be more different to that of the US, something that becomes clearer in the light of the history of the car in Norway.

The History of the Car in Norway

While in the US the car appears through popular culture as the sign of liberalization and the progressive freedom of the individual, tied mainly to the growth of the industrial sector symbolized by Ford on the one hand and Detroit on the other, in Norway the introduction of the car followed a quite different trajectory.

Between 1934 and 1960 import and sale of private cars was strictly regulated ... Each year the authorities established fixed quotas of the maximum number of cars to be sold. To buy a car you needed a permit and to get a permit you needed a good reason. Medical doctors, senior civil servants, and salesmen had no problems getting their permits, while ordinary people would have to wait years to get theirs (Østby 1994: 52 quoting Bjørnland 1989).

This had a profound effect. Without any local automobile industry the car had no relation to the development of commerce, the proletarian work force and liberal freedoms that are seen in the US. In stark contrast these policies made the car a symbol and symptom of the degree of state control and state-led developments of Social Democracy in Norway. At the same time it had the effect of making the car into an item of luxury and individualized privilege. While this met with little opposition in the immediate post-war years, from the 1950s and as a result of growing prosperity, we find a change of mood. Public demand for the car became more forceful and state restrictions were evaded in different ways. A black market prospered and car ownership lost its position as a status object as it became more common. As a result of heavy taxes from 1934–1960, it wasn't until the 1950s–1960s that cars became an accepted part of everyday life. This had wide-ranging effects. As late as 1958, the average Norwegian household spent only 2.5 per cent of their net income on cars, compared to 3.2 per cent on public transport. In 1973, the figures were 14.5 per cent and 3.3 per cent respectively (Sørensen and Sørgaard 1994: 9).

At first glance the next phase following the liberalization of car ownership in 1960 would seem quite different. The period 1958–1973 was one of two that saw a rapid extension of the car into ordinary life; at this stage the percentage of car ownership rose from 6.7 per cent to 19.5 per cent. This remained relatively stable until 1982. As private ownership of cars increased in tandem with greater disposable income, the pattern of living became linked to transportation. Personalized housing in the form of detached single-family houses, which dominates the housing pattern,[4] plus the tendency for larger houses and more dispersed housing patterns, was greatly facilitated and influenced by the availability of the car. The car therefore has had widespread effects in the construction and constitution of modern-day domestic patterns as they have developed since the 1950s and 1960s (Østby 1994).

Østby suggests private ownership was associated with *det gamle samfunnet* or the 'old society' (1995: 7), and with an overclass, characteristic of Norwegian society before the dominance of the Social Democratic Party. It is not surprising therefore that Social Democracy at the peak of its ideological and political dominance acted to control items perceived as luxurious or unnecessary – as the car was perceived to be. In Sweden, however, we find a more nuanced approach. In 1956 the Social Democrat Sven Andersson argued, in his election pamphlet, that the car provided freedom with a new exciting edge, a welcome relief as an instrument both of play and of necessity.[5] The freedom which he propounded as characteristic of the car, however, did not necessarily clash with his political ideology. While the car had been a mark of wealth or class distinction in the past, he argued that the growing frequency of private cars implied that their development complied with the democratization of the time (cf. Sørensen 1992: 40).

During the 1960s and 1970s, the liberalisation of the sale of private cars was the trigger for their growth. During the 1980s, it can be accounted for by the growth in national prosperity. But in terms of the place of the car in the larger political ideology of Norwegian life this is not such a dramatic change. What the car now comes to represent is the increasing success of the Social Democratic consensus that dominated Norway, as in other Scandinavian countries during this period, and with the change in government policies came an indication that the car was beginning to be held as a positive symbol of modern society and proof of a new national prosperity. As we see in a Labour party pamphlet from 1960:

> The decade we are now entering was baptized already at its birth. It was named the golden years. The car and the television stand out as symbols of the new level of prosperity we are now in the process of achieving. (*'Våre oppgaver 1962–65'* – *a note for discussion*, The Labour party, written for the party's annual meeting in 1962, quoted in Østby 1994: 60).

The car became common in tandem with the development of modern Norway, and peaked during the 1980s, a decade often referred to as pivotal in accelerating a recognition of social change (cf. Eriksen 1993). The point here is that once again the car appears as a core symbol of the state itself, in this case a Social Democratic state that claims to have brought affluence and progress while retaining its central role and its commitment to egalitarianism and welfare. While prior to 1960 the state had taken its paternalistic role as its legitimization for severe restrictions on private ownership, in the post-1960 situation it was the 'provider' of affluence and thus access to the car for a much wider segment of the population. The conclusion that emerges from this historical sequence is that from its inception until relatively recently, the car was quite conspicuously a sign of the authority and control of the state both to withhold and also to give to the Norwegian citizen.

This has changed more recently, where national trends have also witnessed a movement away from Social Democracy towards more capitalist privatization, greater disposable wealth and an enhanced self-conscious urban identity. Nevertheless these are only changes of degree. Control by the state may have been taken away when it comes to matters of sales and access, but is still very evident when it comes to the area of consumption and use. There is still a heavy moral emphasis on the superiority of public transport. There are many restrictions concerning factors such as the use of headlights, speed restrictions and control of traffic, all centred around the sense of safety and the prevention of risk-taking. Sørensen and Sørgaard note (1994: 26–7) that this situation seems to be associated with a growth in the statistics of misdemeanours. They also suggest that perhaps 'the role of the car as so to speak instigator of risks and rule-breaking behaviour, may have a very interesting and quite important function as a crystallization point of `immoral' action in a well-regulated and otherwise quite safe society' (ibid.: 27).

The conclusion that they draw is confirmed by the ethnographic evidence for the use of the car as a means to 'flip' over into transgressive behaviour. This takes its specific form against a background of one of the most regulated histories of the car in the industrial world. The car has for its entire history been closely associated with the role of the state, first in terms of the rigid controls that were placed on car ownership, then as an expression of the success story of the Social Democratic political system, and still today in terms of the state's moral and safety concerns, reflected in restrictions of car use. The evidence for this relationship between state control and transgressive con-sumption from the history of the car is reinforced when we examine the parallels in the other medium of transgression, that is alcohol.

A Brief History of Alcohol

Popular moralities which judge the consumption of alcohol in Norway are characterized by drinking on occasions which are strictly delineated from work and everyday life routines (cf. Sørhaug 1996). Alcohol is most commonly employed to mark times as special, such as weekends, festivals, holidays and parties or other special occasions. But when confined to these forums, drinking can be both heavy and almost ritualistic in its departure from the profane routines of daily life. In their report on opinions towards alcohol, Saglie and Nordlund (1993: 18) point out that underpinning the political debate of alcohol in Norway lies the tension between liberalism and restriction, or in other words, balancing individual freedom and collective responsibility. The specific tension which resides between these values in Norwegian mass politics might explain why alcohol policy has been particularly controversial (Saglie 1996: 309). Beer is described as the 'traditional' beverage in Norway until spirits were introduced into the country in the 1600s (NOU 1995: 24). And while spirit production was banned in 1756, the laws relating to alcohol were replaced in 1816 by a liberal political attitude which in part was reacting against former laws imposed by Danish authorities. Spirit consumption increased rapidly, replacing beer and wine consumption, reaching a peak in 1830–40 which remains unparalleled in the country before or after. This dominance of spirit consumption continued until the end of the 1960s, but has now fallen and is on a par with wine consumption (NOU 1995: 24, 18).

It is in the 1820s and 30s that the first restrictive laws were passed, targeting the sale and production of spirits. Two laws passed in 1837 proved pivotal in Norwegian alcohol politics which placed large taxes on the sale of alcohol and allowed for the willingness of individual municipal authorities to sell alcohol in their regions (NOU ibid.). Concomitant with the these came organized teetotalist movements in which 'rural' teetotalism positioned itself against 'urban' standards and practices. The conflict reached its zenith in the period of prohibition from 1916 to 1927 after which time it was voted down in a national referendum (Saglie and Nordlund 1993: 15–16). Teetotalism still exercises considerable influence on public policy and geographically is strongest in the southern and western areas of the country where religious adherence finds its most popular support. Since that time, discussion surrounding alcohol policy remains active as various positions are contested.

In their comparison of alcohol consumption in Britain and Sweden and restrictions placed on its availability by the state, Shanks and Tilley (1987) illustrate some of the issues involved in demarcating heavy drinking as socially deviant. The nineteenth- and twentieth-century middle classes focused particular attention on the 'drinking problem' of the working classes with exaggerated moral purpose, and initiated increasingly restrictive legislative controls between 1850 and 1950, resulting in the gradual and enforced transference of drinking from the public to the private sphere, from the drinking house to the home (1987: 191). Social control and discipline emerge as dominant characteristics of this process in encouraging an efficient working body. While approaches to alcohol consumption and regulation have shifted this century, Shanks and Tilley argue that 'the intervention of the law as a mechanism of control has been displaced by attempts to produce a normalization of the population, both moral and medical' (ibid.: 204). Within the sphere of deviant drinking they argue that in a Swedish context, and contrary to the British experience, alcoholism is framed in terms of ill-health and more specifically as mental illness (ibid.: 205). Increasing proportions of deviant drinkers are to be found in psychiatric hospitals, figures for which dwarf those for Britain. A lapse of state control over the individual is linked to a loss of individual control over oneself. The connections between mental illness and the alcoholic underlines the anti-social quality of this medium, which in itself undercuts popular modes of consumption.

The two main principles in Norwegian alcohol policy are that the state shall have chief responsibility for the importation, production and trade of alcohol – although individual provinces have a certain autonomy in being able to decide what alcoholic drinks it will allow to be sold. Therefore, some provinces, such as along the western and southern belt, might be officially alcohol-free. Other provinces might allow the *sale* of alcohol but not the *serving* of it or conversely the serving of it in public places but not the sale (Hauge 1986: 72). The most important tool for the control of alcohol is the A/S Vinmonopolet or wine-monopoly which was first established in 1922. This state monopoly controls the sale of spirit, wine and strong beers while beers, of a lesser alcohol content are available in supermarkets. Since 1975 advertisements for beverages with a higher alcohol content than 2.5 per cent have been made illegal (Hauge 1986).

In studies conducted by the National Institute for Alcohol and Drug Research, there appears to be a broad support for these measures (Saglie and Nordlund 1993), although in recent decades we can also see a shift in popular attitudes. This shift takes the form of a broad trend towards a more liberal approach to alcohol, coupled with both a reduction of teetotalism and a more relaxed approach to definitions of 'alcohol abuse' (Saglie 1996: 309). Alcohol within a Norwegian setting is approached as a collective medium; it is most usually enjoyed as an appurtenance to socializing with its effects extending beyond the individual in reaching a variety of social spheres. It also carries the stigma of 'contagion'. In view of the fact that nine out of ten bottles of beer are drunk in the company of others, a state agency (*Rusmiddel-etaten*) for alcohol and drug related problems recommends: *If you drink less, the others will also drink less.*[6] Maintenance and loss of control are not only regulated for the individual but are commonly considered in relation to associated social activity. This in turn is reflected in consumption wherein drunkenness, loss of control or a form of abandon is expected, and when this occurs on a group level, ideas of the normative are inverted in a form of group transgression, as in Eva's *vorspiel*. What one does while drunk might bear little correspondence to conventional behavioural norms and for this reason, alcohol is seen as the catalyst which causes this conversion. By controlling alcohol, one is patrolling the barricades which confine the menace of social disorder.

Conclusion

As Sørensen points out, there is, strictly speaking, a history of the car but no sociology (1994: 3). We may study the car in terms of its social effects, but as a sociological unit, the car per se is not seen as being very interesting. Instead, we have to look at cars in terms of networks; household, city, state, culture, etc. In this discussion, I have concentrated on the car and how its significance rebounds from the immediacy of domestic sphere of my informants. But behind the opposition that can be observed through the ethnography between the 'house' and the 'car' lies a more general opposition in which transgressive behaviour is formed within an environment of considerable state control and influence to determine how an object will be viewed. Just as the state is both the paternalistic provider and the presence against which one gets one's kicks, so also the car in addition to its function in executing the daily tasks embroiled in family life can also generate an excitement and the promise of escape from the domestic cares with which it is also associated.

The case is made more clearly when considered alongside the striking parallel with alcohol where the clear demarcation into two opposed forms of behaviour – the sober and the drunk – also seems to link the forms of consumption with a particular history of supply. In both cases a high degree of state control producing an elevated level of normative concern with order is negated by a highly genred 'flip' from one state to another. So it is not just that use of the car by Kari and her friends' appears as particularly transgressive in relation to a Norwegian context, but that its social nature, its gendered nature and its normative nature are all best understood in terms of the particular history of the Norwegian state's attempt to control the relationship between its citizens and the car.

References

Bjørnland, D. (1989), *Vegen og Samfunnet*. Oslo: Vegdirektoratet.

Cohen S. and Taylor, L. (1992), *Escape Attempts: The Theory and Practice of Resistance to Everyday Life*. London: Routledge.

Eriksen, T.H. (1993), *Typisk Norsk: Essays om Kulturen i Norge*. Oslo: C. Huitfeldt Forlag A.S.

Fekjær, H. O. (n.d.), *The Psychology of Getting High*. Chief Medical Officer at Oslo's Agency for Alcohol and Drug problems (*Rusmiddeletaten*) – http://rusdir.no/ SIFA: National Institute for Alcohol and Drug Research.

Gefou-Madianou, D. (ed.) (1992), *Alcohol, Gender and Culture*. London: Routledge.

Giddens, A. (1990), *The Consequences of Modernity*. Cambridge: Polity Press in Association with Blackwell.

Graubard, S. (1986), *Norden – the Passion for Equality*. Oslo: Scandinavian University Press.

Gullestad, M. (1989), 'Small facts and large issues: the anthropology of contemporary Scandinavian society'. *Annual Review of Anthropology* 18:71–93.

—— (1992), *The Art of Social Relations*. Oslo: Scandinavian University Press.

Hagman, O. (1993), 'The Swedishness of cars in Sweden. A study of how central values in Swedish culture are expressed in automobile advertising'. In K.H. Sørensen, *The Car and Its Environments: The Past, Present and Future of the Motor Car in Europe*. Proceedings from the COST A4 workshop in Trondheim, Norway. Published by the European Commission.

Hauge, R. (1986), *Alkoholpolitikken i Norge*. Oslo. Statens Edruskaps-direktorat. 16, 162 SIFA-Rapport.

Hauge, R., Oddvar, A. and Skog O.J. (eds) (1985), *Alkohol i Norge*. Oslo: Universitetsforlaget.

Heath, D.B. and Cooper, A.M. (1981), *Alcohol Use and World Cultures: A Comprehensive Bibliography of Anthropological Sources*. Addiction Research Foundation Bibliographic Series 15, Toronto.

Lamvik, G. (1996), 'A fairy tale on wheels: The car as a vehicle for meaning within a Norwegian subculture'. In K.H. Sørensen and M. Lie (eds) *Making Technology Our Own? Domesticating Technologies into Everyday Life*. Oslo: Scandinavian University Press.

Löfgren, O. (1994), 'Consuming interests.' In J. Friedman (ed.), *Consumption and Identity*. London: Harwood Academic Press.

Nederlid, T. (1991), 'The use of nature as a Norwegian characteristic: myths and reality'. *Ethnologia Scandinavica* 12: 18–26.

O'Dell, T. (1997), *Culture Unbound: Americanization and Everyday Life in Sweden*. Lund: Nordic Academic Press.

Østby, P. (1994), 'Escape from Detroit – the Norwegian conquest of an alien artifact'. In K.H. Sørensen (ed.), *The Car and Its Environments: The Past, Present and Future of the Motor Car in Europe*. Proceedings from the COST A4 workshop in Trondheim, Norway. Published by the European Commission.

—— (1995), *Integreringen Av Bilen i Norge: Forskning og Planlegging*. STS-arbeidsnotat 8/95. Trondheim: Senter for Teknologi og Samfunn.

Peace, A. (1992), 'No fishing without drinking: The construction of social identity in rural Ireland'. In D. Gefou-Madianou (ed.), *Alcohol, Gender and Culture*. London: Routledge.

Rosengren, A. (1994), 'Some Notes on the Male Motoring World in a Swedish Community'. In K. H. Sørensen, *The Car and Its Environments: The Past, Present and Future of the Motor Car in Europe*. Proceedings from the COST A4 workshop in Trondheim, Norway. Published by the European Commission.

Saglie, J. (1996), 'Attitude change and policy decisions: the case of Norwegian Alcohol policy'. *Scandinavian Political Studies* 19, 4: 309–27.

—— and Nordlund, S. (1993), 'Alkoholpolitikken og opinionen'. SIFA -Rapport 3/93. Oslo. *National Institute for Alkohol and Drug Research*.

Sande, A. (1996), 'Rus som sekularisert rituale'. *Norsk Antropologisk Tidsskrift* 4: 302–11.

—— and Henriksen, Ø. (1995), *Rus: Felleskap og Regulering*. Kommuneforlaget.

Shanks M. and Tilley, C. (1987), *Re-constructing Archaeology: Theory and Practice*. London: Routledge.

Sørensen, K.H. (1992), 'Bilen og det moderne Norge: En sosioteknisk transformasjon'. *Tidskrift for Samfunnsforskning* 33: 27–48.

—— (ed.) (1993), *The Car and Its Environments: The Past, Present and Future of the Motor Car in Europe*. Proceedings from the COST A4 workshop in Trondheim, Norway. Published by the European Commission.

—— and Sørgaard, J. (1994), 'Mobility and Modernity. Towards a Sociology of cars'. In K.H. Sørensen (ed.), *The Car and Its Environments: The Past, Present and Future of the Motor Car in Europe*. Proceedings from the COST A4 workshop in Trondheim, Norway. Published by the European Commission.

—— and Lie, M. (eds) (1996), *Making Technology Our Own? Domesticating Technologies into Everyday Life*. Oslo: Scandinavian University Press.

Sørhaug, T. (1996), *Fornuftens Fantasier: Antropologiske Essays om Moderne Livsformer*. Oslo: Universitetsforlaget.

Witoszek, N. (1997), 'Fugitives from utopia'. In Ø. Sørensen and B. Stråth (eds) *The Cultural Construction of Norden*. Oslo: Scandinavian University Press.

Notes

1. The newspaper article was quoting Espen Røysamb from the National Health Board (*Folkhelse*).

2. A little less than £1,000.

3. See Lamvik (1996) on the use and display of American cars (Amcar) in Trondheim.

4. In 1988, 60 per cent of all houses belonged to this category (Sørensen and Sørgaard 1994: 9).

5. *Bilägandet ger fritiden ett nytt spännande innhåll. Man behöver inte vara psykolog för att innse att bilen för en arbetare eller tjänesteman kan bli en välkommen motvikt mot det moderne produktionslivets enahanda. Både som 'leksak' och 'nyttosak' är den svår att ersätta med nogot annat* (cf. Sørensen 1992: 40).

6. *I et 'fuktig' miljø blir det mange stordrikkere. HVIS DU DRIKKER MINDRE, VIL ANDRE OGSÅ DRIKKER MINDRE. I et 'tørrere' miljø blir det færre stordrikkere.*

In a 'wet' environment there are a lot of big drinkers. IF YOU DRINK LESS, THE OTHERS WILL ALSO DRINK LESS. In a 'drier' environment there are fewer big drinkers.

Kwaku's Car: The Struggles and Stories of a Ghanaian Long-Distance Taxi-Driver

Jojada Verrips and Birgit Meyer

19.10.1996. Saturday. This morning we wanted to go to Accra, but on the road to the center I all of a sudden noticed that the temperature of the engine in our car (a Toyota Corolla Estate from 1985) was rising in such an alarming way, that we decided to immediately visit a workshop near our house in Teshie to let mechanics have a look at the problem. Well, they prepared us surprise after surprise. They spent hours trying to solve the mystery of the sudden rise in temperature – to no avail. Not only did the temperature remain disquietingly high, to our amazement and great annoyance our gasoline meter also started to indicate incredible things. The mechanics carelessly removed a particular part (the thermostat) and threw it away – for according to them we would not need it in Ghana anyway – replaced some old parts for new ones, and emptied and filled the radiator that many times that we seriously started worrying about their expertise and fearing for the 'life' of our car. When the friendly smiling gentlemen at last began to remove the radiator in order to clean it, because this surely would end all our technical problems, we told them to immediately stop their seemingly fruitless efforts, for we did not want to have our car completely ruined by them (see Figure 7.1). They are real *bricoleurs*, who try to behave professionally without having the proper tools, but who evidently have no technical knowledge of cars. Pretenders. One just starts with something in the hope that it will work. If you trust these so-called specialists, in no time you will end up with a totally damaged car which hangs together with ironwire and adhesive tape! With each turn of their screwdriver not only its value decreases but also its usability on Ghana's roads, which are not 'car-worthy' anyway (diary fragment Jojada).

Figure 7.1 The mechanics who repaired our Toyota.

21.10.1996. Monday. Today we brought our Toyota to another workplace at Osu (Accra). When we came to collect our vehicle at five o'clock it was not yet ready, though we had been assured that this would be the case. I saw the car standing somewhere with the hood aloft. When I looked under it I got a shock. The radiator was replaced, the original fan removed and a kind of huge propeller attached to the engine block. When I angrily asked a mechanic what the hell had happened, he smiled and told me not to worry at all, because this strange propeller was a 'proper adjustment' to the climate in which our car now had to function. The original fans were too weak for the tropical circumstances, so everybody with a car like ours would sooner or later replace it by a strong one which would always be in operation, and do away with the thermostat, a useless and even dangerous part in these circumstances. We felt horrible and irritated, for once again the value of our car had gone down due to this so-called adjustment. The mechanic, however, tried to explain that we should be glad with it, for now our car was 'tropicalized' or 'baptized into the system', but we could not be glad at all. These people not only don't ask you anything, they also devaluate your property because they are *bricoleurs* and no real mechanics who respect an engine (diary fragment Jojada).

These two fragments from Jojada's diary show how shocked we were by the way in which Ghanaian car mechanics treated our more-than-ten-year-old second-hand car (for which we had just paid 7.6 Million Cedis and which was praised by our Ghanaian friends and neighbours for its newness and beauty) when it had some minor technical problems.[1] These people really made us angry by their seemingly respectless and unprofessional approach towards the engine. However, as we eventually discovered, this approach was not an indication of their lack of technical knowledge, and our anger rather indicated our own ignorance and bias about cars, their upkeep and repair.

We visited Ghana, in the case of Birgit, to study Ghana's vibrant videofilm industry and, in the case of Jojada, to continue research on the artisanal fishery sector, and, like many anthropologists working in urban areas, had just bought a car in order to drive around. While initially we took the use of a seemingly familiar thing as a car for granted in the same way as we did at home, our experience with mechanics made us understand that cars were an incredibly interesting, yet unduly neglected research topic. For that reason, Jojada gradually shifted his attention away from canoes and started a small odyssey through the realm of cars in Ghana. Though not focusing on cars directly, Birgit also realized the importance of car matters, both in films and for their audiences, and collected information about this. The motive behind these endeavours was to deal with our first, strongly negative impressions of car mechanics and electricians, and to give room to a gnawing feeling that these impressions were totally off the mark and appeared to reveal more about a Western fetishization of cars than about actual Ghanaian expertise with regard to vehicles. In order to understand the importance and meaning of cars in contemporary Ghana, we sought to find out how one obtains them, uses them and keeps them rolling on the often defective roads, how they are 'adjusted' or 'tropicalized,' how one repairs them with a minimum of tools and an often discouraging lack of spare parts, and what kind of perceptions of and stories about cars were circulating.

Our point of departure became a car which had appeared in the open space next to our house in Teshie just a few days before our car troubles started. This Peugeot 504, which looked like a heap of scrap-iron beyond repair, was nevertheless day in day out patched up by a young man and a few assistants. At the time we developed an interest in this vehicle, it was nothing more than a beautifully blue-sprayed body with the slogan *God Never Fails* on its back and a worn-out engine inside. To mention just a few glaring defects, it had no electrical wiring, the coverings of the doors and the seats seemed to have disappeared, the tyres were in a terrible condition and the exhaustion pipe almost fell on the ground (see Figure 7.2). The man who invested so much time,

Figure 7.2 The interior of the 'God Never Fails' before it was repaired.

energy and money turned out to be long-distance taxi-driver Kwaku A.[2] who together with his young wife and baby lived in two rented rooms nearby. Several months before, he had had a serious accident with his car and now he was trying to bring it back on the road again, for he desperately needed some income. Soon after we got in touch with Kwaku, we found out that the Peugeot 504 had been imported from the Netherlands in September 1991. To our surprise, we realized that it had crossed our path in Ghana before; we even had made a ride in it in the company of one of its former owners, a Dutch development worker whom we met during an earlier trip to Ghana. This Dutchman sold it somewhat later to a befriended South African doctor, who in his turn sold it to Kwaku for the amount of $3000 in April 1994, on the condition that during one year he was to pay $250 per month to a certain Ghanaian woman.

This rather remarkable coincidence prompted us to write at least part of the 'biography' of Kwaku's car and to use it as an empirical entrance into the world of cars in Ghana. Soon after we met Kwaku, we therefore made an agreement with him: in exchange for all kinds of data, he would take Jojada on small trips and introduce him to relevant people and places to invest in the reconstruction of his beloved car, the *God Never Fails*,[3] so that he could start as a long-distance taxi-driver again. Kwaku initiated Jojada in the scarcely explored, thoroughly male domain of the car world in Ghana.

Given that all over the globe cars play a tremendously important role as means of transport and sources of power and prestige, it is surprising that they are so much neglected by anthropologists. When we started looking for anthropological work on cars and car-related matters, we realized that the few existing publications focus on very partial aspects. There is, for instance, some anthropological literature on drivers (with regard to America see Agar 1986; with regard to Africa see Field 1960; Jordan 1978; Peace 1988). These authors, however, merely focus on social networks, contacts with passengers and other sociological topics, and pay virtually no attention to the car in its materiality. The same applies to anthropological studies of car slogans (e.g. Lawuyi 1988, Van der Geest 1989), which regard the mottos car owners chose to paint on their cars as a reflection to their world view. In his chapter on a ride in a Songhay bush taxi, Stoller, too, confines himself to a Geertzian symbolic analysis of 'the complex of Songhay bush taxi interaction' (1989: 69–84). We do not, of course, wish to deny the merits of approaching cars as vehicles of meaning (see also White 1993). But our point is that research should not be limited to symbolical approaches but should also address the more mundane and material aspects of cars in their use in everyday life.

It seems that, up until now, Kopytoff has pleaded in vain for more anthropological research on cars, conducted along the following lines:

The biography of a car in Africa would reveal a wealth of cultural data: the way it was acquired, how and from whom the money was assembled to pay for it, the relationship of the seller to the buyer, the uses to which the car is regularly put, the identity of its most frequent passengers and of those who borrow it, the frequency of borrowing, the garages to which it is taken and the owner's relation to the mechanics, the movement of the car from hand to hand over the years, and in the end, when the car collapses, the final disposition of its remains. All of these details would reveal an entirely different biography from that of a middle-class American, or Navajo, or French peasant car (1986: 67).

It would lead too far to discuss the reasons for anthropologists' reluctance to write such 'auto'-biographies.[4] Certainly Otnes, in his reflection about the fact that social scientists pay so little attention to cars as objects,[5] hits an important point:

A general reason . . . is quite simply that much of sociological theory . . . has a tacit but nevertheless a strong anti-materialist bias in which an artifact becomes a 'mere vehicle,' a mundane means of little substance or interest (1994: 50).

Next to this general anti-materialist bias, which has been opened up in anthropology only in the course of the last decade (e.g. Miller 1987, 1995), one may also notice a bias against modern technology. Resonating with a mistrust of technology propounded by Western intellectual tradition (expressed most explicitly by thinkers such as Max Weber and Martin Heidegger), in Africanist anthropology there still is a tendency of favouring the study of local, cultural traditions. While it may no longer be stated overtly that African drivers show a 'lack of "feeling" for machines', and that their allegedly defective attitude towards cars 'may be the result of a youth passed in a non-mechanic world' (Morgan and Pugh 1969: 608, quoted approvingly by Jordan 1978: 41), the view that Western technologies are foreign to and potentially disruptive of African culture is still lingering on.[6]

Paradoxically, in the practice of everyday life in Western societies, this anti-technological bias is accompanied by an uncritical attitude which just takes technology in general, and the car in particular, for granted. As the Dutch sociologist Van de Braak (1991: 50ff) has argued, there is a rift between production and use of cars. Consumers simply expect that a car keeps on rolling; only at the very moment when something is wrong one realizes how little one knows about car technology and quickly seeks help from a specialist, who, in turn, is expected to replace the necessary parts without changing the technological structure as such. This unquestioning stance, masked by blind confidence, can best be characterized as alienation (ibid.: 51).

In this chapter we seek to overcome the anti-material and anti-technological bias of our discipline, as well as the typically Western, taken-for-granted attitude towards cars which we appeared to reproduce in our reaction to the Ghanaian mechanics, and investigate cars in their mundane as well as symbolical, material-technological as well as spiritual dimensions. The foremost goal of this chapter, then, is to understand, by way of a detailed study of Kwaku's car, why and how cars in Ghana are culturally redefined or, as Ghanaians put it, 'baptized into the system,' 'tropicalized' or 'adjusted,' and to investigate how cars are kept in the system for periods Westerners usually deem technically impossible. On the basis of our – necessarily tentative – answers to these intriguing questions, we will briefly reflect on the car as a marker, and indeed, a vehicle of modernity. Before dealing with the vicissitudes of the Peugeot 504 it is necessary, however, to say something in general about Ghana's infrastructure, car trade, the Ministry of Transport and Communications, the insurance business, types of car use and (police) control.

Roads, Customs and Licences

Ghana is almost fully dependent on lorries and cars for the transport of goods and people, for it does not have an elaborate railway system.[7] Ghana's road network is most dense in the southern regions and least in the northern ones. Due to the bad condition of these roads, the reachability of particular places and areas often forms a serious problem. Highways exist only in, around and between big cities such as Accra and Tema. The rest consists of asphalted interregional two-lane roads, untarred country roads and bush tracks. The problem with many of these roads is that they are full of potholes, and therefore rather dangerous, especially in the dark, for outside the big cities lighting is almost always absent. Nowadays the government invests a lot in the maintenance and improvement of Ghana's road network, but it has not yet been able to bring the quality of the network to an acceptable level which matches its intensive use. For many Ghanaians, this is a constant point of criticisms with regard to the state, which is regarded as unable to cater for the needs of drivers and travellers. As Ghanaian popular videomovies[8] indicate over and over again, good roads and beautiful cars are much cherished and figure as ultimate icons of modernity. There are endless scenes in which the camera takes the audiences on a mimetic ride through the beautiful highways and lanes of Accra – scenes which prompt audiences to loudly admire the beauty of the capital city, which eludes them in their everyday life, and which apparently can only be seen from the perspective of a smoothly-running limousine.

As Ghana does not have its own automobile industry, the country is fully dependent on the import of new and second-hand cars, trucks, lorries and busses from abroad. Until a few years ago, the import of used vehicles from European countries, for instance the Netherlands and Germany, was flourishing.[9] Lots of old vehicles, sometimes declared unroadworthy in these countries, were shipped to Tema where they could be collected by their owners (private persons, car dealers, or companies) after they had paid a certain amount of customs duties, including a so-called over-age tax. The latter kind of tax is considerable when a car is older than a particular number of years. One of the tricky things with the import of old cars is that the government tends, on the one hand, towards decreasing the age beyond which this tax has to be paid and, on the other, to increase this tax each year. The background of this policy is to muzzle the import of too many cars above a certain age.[10] For the Toyota Corolla we bought in 1996, the trader who imported it had to pay 50 per cent of the C.I.F. Value, that is C 3.767.327.00, as over-age tax plus 10 per cent (or C 753.466.00) customs duty. Although a price of 7.6 Million Cedis for a more than ten year old car that would cost no more than $500 in Europe may seem excessive, if we add to this the amount paid for it abroad, the costs of shipment and insurance, then it is understandable.

Against the background of such high prices for used vehicles, it is no wonder that people struggle to keep them on the road as long as possible. Many people get into serious debt with relatives and friends in order to get their certificate of payment of custom duties in Tema harbour, without which certificate registration by the *Vehicle Examination and Licensing Division* is simply impossible. These problems frequently occur in the case of cars shipped by ex-patriates to relatives in Ghana, or cars bought on trips abroad by Ghanaians without sufficient knowledge of the latest tax rules and regulations.

After an imported car has been released, more obstacles have to be overcome before one can finally drive it. Officially, a vehicle has to be technically examined by state officials, for without such an examination one will not get a licence number. Getting such a number may be a difficult affair, if one does not know the right official and/or one has not enough money to ease one's way through the bureaucratic channels. Therefore car owners often let others go through the time- and money-consuming procedures at the licensing offices. In and around these offices are found many 'agents' – who make a living by offering their services in getting the proper registration documents within reasonable time. As in the Cameroonian situation described by Mbembe and

Roitman (1996), here too the official bureaucracy is seconded by its 'fake parallel', and only through particular 'ways of doing' can one access the necessary documents. The amount of money agents demand for their knowledge of the proper procedures and how to circumvent them, for instance by offering officials they know (or work with) small 'gifts', can be rather high, but worth the cost, especially if one is in a hurry. A government decree in 1996, that every car owner had to re-register his or her vehicle before a particular date, resulted in a bonanza not only for these agents but also a host of others, such as the insurance people, the producers of licence plates, the sellers of all kinds of car accessories, the owners of small repair-shops and vulcanizers, whom one can regularly find in the direct vicinity of a licensing office. The final step in obtaining the formal right to drive the car after the certificate of roadworthiness and a licence number are obtained is to insure one's vehicle. Even without the documents mentioned, the police may be lenient in exchange for 'presents'. So the 'adjustment' of the car as a mechanical process is accompanied by similar adjustments in the cultural norms of Ghanaian bureaucracy and commerce.

Public Transport

When one wants to travel in Ghana, one travels by car. Though the state maintains particular bus lines between big cities with rather good material, this kind of public transport is of minor importance compared to the private sector. Ghanaians travel within and between cities with private taxis and *trotros* (derived from the three-pence charge one once had to pay for a ride). Formerly these *trotros* or *Mammy lorries* were Bedfords, which were imported in great numbers from Britain as complete vehicles between 1948 and 1959 and thereafter C.K.D, that is 'Completely Knocked Down', to be assembled in Tema until 1966, when this type of lorry was banned (cf. Kyei and Schreckenbach 1975: 8). These characteristic vehicles, almost all with evocative slogans in the front and at the back, which for more than three decades dominated the image of traffic flow on Ghana's roads, packed with passengers and all kinds of goods such as yams, fish, fowl and fruits have almost all disappeared now. Nowadays, the transport of persons and their goods is mainly by passenger cars and station wagons made into short- and long-distance taxis, as well as with all kinds of delivery vans, ingeniously transformed into small busses, and ordinary busses. The passenger cars are painted yellow at the front and back of each side and have a small black shield which indicates the name of the car owner, the maximum amount of passengers, etc. The long-distance cabs (mostly Peugeot station wagons, but also small and big busses) normally also have such a shield, but they are not painted in a specific manner. Instead they almost always have painted slogans, such as *Sea never Dry, Don't Kiss the Horror* or *God Never Fails* (cf. Kyei and Schreckenbach 1975; Date-Bah 1980; Van der Geest 1989).[11]

In a big city like Accra the apparent chaos hides an order in which many drivers work in turns on particular trajectories between stations, use fixed charges and belong to unions with elaborate rules and regulations. The unions often run (rented or owned) stations[12] and supervise the admission of new taxi-drivers on the basis of such criteria as experience, knowledge of a trajectory, state of the vehicle and the possession of the proper documents. After having been accepted as a new member, one has to pay income and welfare tax as well as booking and union fees. Not all taxi-, *trotro*- and big bus-drivers own their vehicles. Many of them drive for relatives or other private persons, who often live abroad. It is exactly this practice which forms a frequent source of troubles. Common complaints by the owners, for instance, are that their drivers do not maintain the vehicles properly and, still worse, do not stick to the financial agreements made with them but instead keep too much money for themselves. Conversely, drivers complain about owners' greediness, their inclination to attribute any technical fault to the driver, and their general indifference towards the hardships involved in the job.[13] These potential conflicts notwithstanding, driving a taxi or *trotro* is considered a respectable business, which not only conveys prestige, but often earns more money than one can get through other jobs. Therefore becoming a driver, and possibly a car owner, is aspired to by many young men.[14] Our neighbour Kwaku was one of them, and he even managed to get his own car exceptionally fast. His case shows that the high hopes vested in vehicles do not always materialize and that driving an old Peugeot between Accra and Takoradi may be a heavy, almost unbearable burden.

Struggling with 'God Never Fails'

Kwaku was born in Teshie,[15] where his father (a Ga) worked as carpenter. As he did not like to stay in the family house after his mother's (a Fanti) death, he ran away from home at the age of 13; first to Labadi (the neigbouring Ga village which has became absorbed by Accra) where he became a driver's mate, and later to Takoradi (Western Region) where he started living with his mother's sister's husband. After a few years he returned to Labadi where he became a mechanic's apprentice, for he loved cars and later wanted to have his own workshop, but the buying of tools turned out to be an unsurmountable obstacle.[16] He got his driving licence and earned some money by driving around his elder brother, a rather successful businessman. Then he obtained a job as a driver for a hotel owner in Accra. While working for this man in the early 1990s, he became acquainted with the South African doctor Henry H. and his Dutch friend Klaas W., the development worker. Through the doctor he obtained his first taxi, a Datsun, which he bought for C 457.000, a rather low price at the time (September 1992). Since he only possessed C 107.000, Kwaku had to borrow the rest from his friend, whom he very quickly paid back in monthly installments of C 25.000.[17] Very soon after he got it, Kwaku gave the car to a driver to exploit it for him in the morning; he himself used it during the night. Because of troubles in the hotel – the owner and his wife did not understand where Kwaku's money came from and accused him of theft – he left the place for the doctor's house and became his 'houseboy'. Though Kwaku invested much money in the maintenance of the old Datsun, he still earned a lot with its exploitation, at least enough 'for the house' and for his expensive marriage with Rose in March 1994.

He managed to organize a wedding many young people dream of (and which forms a recurring, characteristic feature of popular video movies): the bride was wearing a white gown, the ceremony took place in church, and there was a reception afterwards. Through this church wedding – in the *Apostolic Church of Ghana* to which Rose and her family belong – the marriage became officially registered. Many young women want to convince their boy friends to make such an arrangement, which is also much favoured by Ghana's booming pentecostal-charismatic churches (cf. Meyer 1998), because official registration makes it legally impossible for the husband to take a second wife. As Mr and Mrs, Kwaku and Rose commanded much respect from neighbours when they moved to their beautifully painted two-room apartment in Teshie: a modern, Christian couple with a comparatively fine income. While Kwaku was very proud of having been able to marry in such a way, he also felt that many people, especially in his own family, would envy them and seek to spiritually destroy their happiness.

And indeed, soon after the wedding, Kwaku got into trouble, when for unknown reasons the car burnt out. He managed, however, to sell the heavily damaged vehicle for a higher price than he bought it for. In April 1994 his employer and friend Henry sold the Peugeot 504 he had bought from the Dutch development worker to Kwaku. This time the price was considerably higher, that is $3000, to be paid in monthly installments of $250 into the bank account of the doctor's Ghanaian girlfriend, who was to remain the official car owner until the payment had been settled.

Kwaku started to invest in the car immediately after he got it. He transformed it into a comfortable long-distance taxi with which he began to transport passengers and goods between Accra and Takoradi. His home base became a station in Kaneshie owned by a local co-operation (the Kaneshie Co-operative Association) where he joined an independent local group of Peugeot drivers (not yet resorting under the Ghana Private Road Transport Union [GPRTU] of the Trade Union Congress [TUC]) working on this lucrative trajectory. In order to be able to maintain his car well, he first bought the body of an old Peugeot 504 for C 200.000, and somewhat later also the engine for C 500.000. He left both in Takoradi to cannibalize them just in case the need might arise. This is a common practice among long-distance drivers who mostly work with rather old and worn-out material.

For ten months Kwaku drove without many problems and therefore earned a decent living and had a good time with his then pregnant wife.[18] But then an endless series of troubles with the car started, which eventually made Kwaku decide to become a good Christian again, for these problems almost ruined their life. It all started in February 1995 with three flat tires in a row, each at a place where there was water at both sides of the road – incidents which worried Kwaku a lot, for 'this could be no coincidence'. He visited a church elder, an old lady, who told him to fast for seven days. Yet he did not stick to her advice and on April 13, on his way back to Accra, he had the first serious accident with his Peugeot near Malam Junction not far from Takoradi. He needed almost three months to get the spoiled car back on the road again. Though the insurance company paid him C 500.000, he still needed to borrow money from relatives. Fortunately the doctor, who had left Ghana for some time, appeared on the scene again. In August he hired Kwaku and his car to travel, and this reduced Kwaku's debt to 1.2 million Cedis.

But from then on the car became a real nuisance, a money-devouring monster. Kwaku's diary in which he meticulously recorded any event related to his car – his auto-biography, so to speak – contains entry after entry on disturbingly knocking bearings, malfunctioning cylinders, an out-of-order crankshaft, a leaking radiator, etc. Time and again he had to make embarrassing trips to uncles in order to borrow money from them, which they sometimes refused. He even considered meeting the church elder again to pray for him, so that his predicament might end. In spite of all these technical problems, Kwaku managed to pay off part of his debt. On April 13, exactly one year after his first accident, he had a serious second one.

> I thought that I and my passengers would die. I really felt there is something behind, so that I cannot pay the money, so that she will get the car. So let us pray over things and stop joking. (. . .) I was reporting to her that the car was not working properly and asked her to give me some money. She said I had to go and come. I did that three times, but she did not give me the money I asked for. Then I decided to borrow the money elsewhere. I have not been at her place anymore. (. . .) The thing has become a sort of pressure on me. I want to get rid of Mary and transfer the car on my own name.

From then on Kwaku did not pay Mrs Mary P. for a long time. At the time we met him in September 1996, he still owed her C 600.000; and had stopped paying her since that April after he had had his second serious accident with the car. He believed that she was behind it – that is, that she had used some destructive 'spiritual' means – because she wanted to get the vehicle back. Since his car was seriously damaged and moreover showed more technical shortages with every day, he decided to stop making trips to Takoradi and back and to invest all the money he could borrow in slowly repairing his vehicle. He even used the capital (C 200.000) his wife had accumulated in order to start a trade. This, of course, put tremendous pressure on their relationship, especially since Rose had become pregnant again. They had frequent quarrels, and not much was left of the image of the modern, better-off couple they had given before. Rose often stayed with her own parents, and expected Kwaku to become a better husband – a project he himself assented to in discussion but could not always live up to because of his allegedly quick temper.

Kwaku now became a regular visitor of repairshops in Teshie and all over Accra. He cleverly used his elaborate network to recruit a whole series of specialists, from sprayers to electricians, who could help him in bringing his car back to life again. Sometimes, however, these people caused more problems than they solved. Once a welder worked so carelessly that the electric wiring of the Peugeot caught fire and was partly destroyed. This is no small wonder, for welders, who usually work without protecting their eyes, are not used to removing inflammable material in their direct environment. When something happens they simply use a small plastic bottle with water to extinguish a fire (see Figure 7.3). In this case to no avail, with the result of a damage of C 60.000 of which the unlucky welder could only pay C 20.000 immediately.

Figure 7.3 A welder at work.

At the time we developed a serious interest in Kwaku's car, he and a befriended electrician were busy installing new wiring (see Figure 7.4). But they could not properly test it, for the Peugeot's battery urgently needed to be repaired or, still better, replaced. The problem was not to

Figure 7.4 New wiring for the 'God Never Fails'.

find a battery repairshop, which abound in Accra, but the money needed to repair it. In these shops, one repairs, refills and recharges for a price dependent upon the number of plates in a battery.[19] It is here that our financial involvement in the revival of Kwaku's car began, for we decided to help him buy a new battery in exchange for information. But this proved hopeless, for the vehicle was in bad shape. To list only a few of the technical problems, the shock-absorbers were weak and needed to be refilled with oil, the radiator and the fuel pump were leaking, the bearings were knocking, the cylinders, pistons and crank-shaft were worn out, the starter often did not work, the fanbelt was damaged, the gearbox looked dangerous, no meter at the dashboard was working and almost all of the fuses were fixed instead of replaced by new ones.[20] As Kwaku used to say: 'The body is nice now, but the inside is not good.'

In order to make the inside better, so that he could start his business again, he did what so many of his colleagues would do in similar circumstances. He fell back on using all kinds of locally invented 'adjustments' and, when these would no longer work, on buying second-hand or even new spare parts. Since original parts are often unavailable in Ghana or much too expensive, one frequently makes do with cheap imitations of an inferior quality produced in countries such as Nigeria and India. Since even these copies are rather expensive, the common practice, however, is to first pay a visit to one of Accra's many markets, for example at Kaneshie, where old engines are cannibal-ized and a wide variety of second-hand parts are offered. In these areas one can also find a host of small workshops where old parts, which are usually thrown away in Western countries, are most ingeniously 'revitalized' or utilized to repair engines. In the machine-shop *Godfirst Engineering Works*, for instance, the walls of old cylinders are turned into 'new' ones to be placed in engine blocks with worn-out linings (see Figure 7.5). Here one is confronted with a tremendously rich sort of technical knowledge and practice, which came into existence due to a shortage of money and hence of means and parts. This is backed by an impressive degree of inventiveness.[21]

Kwaku is a master in using technical 'adjustments'.[22] For example, he replaced the two-chamber carburettor of his old Peugeot by a one-chamber one in order to save petrol. The air inlet was not original any longer, but consisted of a piece of pierced tin. A lot of gaskets and rubber parts, such as bushings, were 'indigenous', that is, cut out of old tubes and tyres (see Figure 7.6). The art of recycling these materials is practised with a minimum of tools by a small army of highly skilled

Figure 7.5 The production of new linings in the machineshop *Godfirst Engineering Works.*

masters and their apprentices. All the fuses in the Peugeot were repaired with copper wire, because original ones could not be found or were too expensive. At several places nails were used as lock pins. Some rubber tubes were fixed with iron wire, whereas others which evidently were out of use were closed with old spark-plugs, butterfly nuts or even pieces of wood.

Figure 7.6 A steering rod with a rubber 'adjustment'.

Next to these directly visible 'adjustments', there were a series of hidden ones. In order to prevent the knocking of some worn-out main bearings in the engine block, Kwaku had put pieces of greasy paper between them and the crank-shaft. Instead of using a special spring, his distribution chain was held in its proper place with a piece of copper pipe. He had raised the oil level in his shock-absorbers beyond normal so that he could drive more comfortably on roads full of potholes. He more than once asserted that in case of an emergency situation, he would not hesitate to temporarily use soap-suds instead of brake fluid.

A more careful inspection of *God Never Fails* would certainly have revealed more technical 'adjustments'. Similar ones and others, such as the huge propeller-like fan put in our car, could be observed in other vehicles.

It is important to realize that this widespread 'tropicalization' of motor-cars in Ghana (as well as in many other African countries) not only rests on a thorough knowledge of how engines work, but also and especially on a rather unique type of knowledge of how one can keep old ones working in a situation of limited goods. People like Kwaku not only have to cope with a shortage or lack of spare parts, they often do not have the proper tools to repair their vehicles, and for that reason they often involve their bodies in ways that compensate for the lack of such tools. For Kwaku it was almost normal to suck petrol out of his tank in order to clean certain car parts or to use his tongue to feel if his battery was still charged. And since the gasoline, oil and temperature gadgets had stopped working long ago, on his trips between Accra and Takoradi he constantly tried to figure out by listening and smelling, whether his car was still going strong.

Only a few garages in Accra are equipped with sophisticated instruments to repair cars, for example those of big dealers, but they are too expensive for ordinary people like Kwaku. In contrast to the often poorly equipped workshops of many mechanics[23] – who like to advertise themselves as, for instance, 'gearbox doctors' – these garages look like sophisticated 'hospitals' for machines. It would be wrong, however, to presume that the workshop mechanics on the one hand, and the employees of these modern 'hospitals' on the other, represent respectively an inferior and a superior approach to the car. They share a thorough knowledge of how this pre-eminent specimen of modern technology works and best can be kept working under the given circumstances, but differ in the way they are able to treat it when it refuses to function properly. At this point mechanics and people like Kwaku show an incredible inventiveness based on a perception of the car, more especially the engine, as a thing or an object which technically can be dominated and domesticated. And they need this, for old cars constantly confront their users with new problems to be solved.

Kwaku, for instance, managed to get his car back on the road in the middle of November, but was faced with a serious breakdown of his engine less than a month later. On a trip from Accra to Takoradi his crankshaft became loose, as a consequence of which at least one connecting rod of a piston was totally spoiled and others seriously damaged. In the meantime he had earned a little bit of money, but not enough to easily overcome this new disaster. We left Ghana when he was struggling to find an additional sum of money to the one we 'borrowed' him. But he was successful, for in 1999 he was still on the road between Accra and Takoradi with *God Never Fails*, this time sprayed yellow, and had managed to pay off his debt to Mary and finally become the official owner (see Figure 7.7).

Figure 7.7 Kwaku and his yellow-sprayed 'God Never fails' in 1999.

Though it seems as if mechanics and people like Kwaku 'will fix everything and bring it back to life again', this does not mean that their perception of cars, especially engines, and how they (dys)function is merely technical. Here one enters the fascinating area of the beliefs, images and lore with regard to these material objects. Just like the canoes used in the artisanal fisheries, they very often are seen and treated as a kind of beings with a will of their own, who can get hurt and die and therefore have to be carefully protected against all kinds of evil influences. To make things more complex, most drivers, as is also the case in Western societies, tend towards a strong identification with their vehicles, that is, with already semi-anthropomorphized material objects (cf. Verrips 1994).[24] At times they feel so much one with their car, that what happens to these 'beings' also happens to them. And though drivers know that many cars are old and therefore run more risks of collapsing than new ones, that accidents occur as a consequence of bad road conditions, overloading, speeding and the use of alcohol,[25] they nevertheless want to have an answer to such pressing, classical questions as to why a car collapses or gets involved in an accident at a particular time and place, so that income is lost.

We will start by describing some of the practices or rituals to protect cars from collapsing and getting involved in accidents and then present some of Kwaku's stories about why such things happen. Apart from painting slogans of the type *God Never Fails* on their cars, drivers often also use stickers saying, for instance, *I am covered by the blood of Jesus* or *I will make it in Jesus name*. Next to this kind of protective formulas on and in their cars, Christian drivers often ask their pastor and church elders to pray over their vehicles, so that they will be protected against the influence of evil forces or Satan. Muslim drivers also visit their malams to pray over their cars, so that they will be protected against evil. Some of them even go so far as to park the car in front of a malam's house so that it will not be stolen.[26]

Kwaku himself went through a protective Christian ritual just before he started making long distance trips between Accra and Takoradi again: '. . . we prayed that whatever ties there were on the car with evil, God should burn them. Seeking for the blood of Jesus Christ.' When he was about to take his car back on the road in November 1996, he invited us to accompany him to the *Apostolic Church* one Sunday. In order to mark his good intentions and high hopes for the future, he attended church together with Rose after quite a long period of absence. After service the pastor and a number of church elders blessed the car through prayers (see Figure 7.8). When we told him in the course of our discussion about this ritual that we did not attend church at home, Kwaku said that he could appreciate that, as he himself had not been a church-goer in the past. 'But', he said, 'you don't have *juju* [black magic]. But we here, we have it here, and we have to protect ourselves against it and the only way to do it is through the Holy Spirit.' Clearly, Christianity is regarded as generating power and protecting people against evil forces.

Kwaku, however, did not assume that such rituals would protect him against all kinds of troubles, for he explained:

> It is protected, but not that nothing will happen to it. Against evil forces and unwanted accidents, but not against technical failure. The car is old. Technical failures are caused by lack of maintenance. It is essential that

Figure 7.8 The blessing of Kwaku's car in 1996.

you grease it. It is like our own bodies, you have to put grease in the joints. If the pastor prayed over it, there still should be oil, water, all what is necessary. Like a body you have to feed it.

This statement notwithstanding, he often talked about evil forces spoiling totally worn-out parts of his car. For him in everyday life technical and supernaturally caused failures were entangled, so much so that a clear-cut boundary between the two domains could not be drawn. In any case, he started each journey with a silent prayer because he believed that regular praying and listening to what God said to him while under way were important ways to avoid all sorts of trouble. He was also convinced that living according to Christians rules might yield luck on the road and a decent income.

Kwaku told us many stories about the use of *juju* to protect cars and drivers. According to him, there were many non-Christian – but also 'nominally' Christian – (taxi)drivers who had hidden special things, such as particular beads or pieces of (blood-stained) cloth, under the hood or the pedals, in the steering wheel or even the engine in order to drive safely and earn a lot of money with their trade. He especially feared (truck) drivers from Nzima in the Western region, for some of them were supposed to have very strong magical powers

> . . . which make that, if the car is gonna crash you might not find them anymore, the drivers. They might be somewhere else. (. . .) It is just like an air bag. As soon as you get the impact, the airbag will explode to the one steering. That's how it is, as soon as the accident comes, he vanishes.

But according to Kwaku 'most of the drivers have got a kind of protection which they might not tell. Not for vanishing or disappearing, but sometimes it protects them from evil forces.' In spite of all these protective measures, drivers every now and then are confronted with these forces and their destructive influence. They might manifest themselves, for example, in suddenly broken engines, mysterious flat tyres, and more or less serious accidents. In Kwaku's view, and much in line with the stance of the pentecostal-charismatic churches, all these evil forces come from the realm of the Devil, who was sacked from heaven by God and dragged along all demons (cf. Meyer 1999b), and are called upon by envious or greedy persons to afflict cars and their owners. There are many stories circulating around spiritual pacts made between local priests and their clients, which involve killing others spiritually through car accidents, or even offering a whole car as a blood sacrifice in exchange for personal gain (cf. Meyer 1995). Other stories are about strange beings roaming about in the night, and seeking to distract drivers' attention so that their car will perish. Kwaku told us that he had often seen such beings along the roadside, but that as a result of his faith in God he had always managed to get through.

Due to the presence of witches among the passengers, a car may become 'strong', 'stiff' or even stop all of a sudden at a certain place, so drivers believe.[27] Kwaku experienced this several times; once, for instance, when he had Hausa people from Ivory Coast in his car who carried cola and a strange bag with them:

> I was coming from Takoradi and as soon as I got to the big bridge, the car couldn't go anymore. I turned back, then I drove fine. I returned back again, it couldn't go again. So I drove to the station and dropped them to the next car. (. . .) I didn't understand it. In certain traditions cola means something. Maybe they are having something of a dead person, the hair or the nails. If your car is not that type of car, it can't go.

There are also many stories about the ghosts of persons who either died in a car or were transported in it to be buried, which prevent automobiles from functioning properly. Kwaku told us, for instance, how the ghost of his deceased stepmother stopped the very small car in which he and some of his relatives were travelling to her burial.

> We saw oil coming out, smoke, so we stopped. The driver was complaining 'why?', nothing was wrong with the car. The woman wanted us to be comfortable and knew the car that they brought was very big, a lot of space. So the body was put into that car, then we had more space. It was two hours to the village where we were going, in the Volta Region.

In order to prevent this kind of trouble, one often pours libation at mortuaries when collecting a corpse and ritually purifies cars in which people died by accident.[28]

But it is not only envious outsiders who may cause technical breakdowns and accidents. Kwaku, for instance, was afraid that jealous uncles might poison him and spoil his car because of his relative success as a long distance taxi-driver. He also suspected one of his father's wives of seeking to spiritually spoil his business, because she was jealous that he – 'just a small boy' – could own a car. In dreams and visions, he every now and then 'saw' how envious 'witches' used his car spiritually, and the next day he felt that the car was tired and run-down. It was, however, very difficult to recognize them, for in his visions – about which his pastor and a prophet said that they were justified – they were using different people's faces. Nevertheless he was pretty sure that they were close relatives who wanted to destroy him and his car. His experiences with witchcraft – here, too, the 'dark side of kinship' (Geschiere 1994) in the sense that he saw himself as a target of his relatives' envy – formed one of the reasons why he would not like to live in the family house and felt in constant need of protection.

Conclusion

By now it may be clear that our initial anger and amazement about the treatment of our Toyota, which we found increasingly embarrassing as our research proceeded, reflects a particular Western way of dealing with cars. At least in our experience at home, car repair is a specialized yet centralized affair. Like many car owners, we take the technological dimension of cars for granted and simply trust that, regular maintenance at the garage provided, everything works. In case something might go wrong, we would never dare to touch the engine of our VW ourselves and rather leave it to the care of expensive specialists, who do not admit snoopers at the workplace. And, afraid to get stranded somewhere along the road, we are prepared to follow the mechanics' advice to replace all sorts of parts even before they break down, or, alternatively to buy a newer car when maintenance of the old one becomes too expensive. All this pinpoints a rather ignorant and, in a sense, alienated attitude towards car technology.

How different are things in Ghana where there are great numbers of different specialists for various aspects of cars, where parts will only be replaced after they broke down, and where even lay people have an admirable working knowledge of motors and can easily engage in technical debates with the mechanics. The distance between the realms of repair and use, so characteristic of Western societies, is blurred. On the whole, it appears that the engine commands much less awe. People easily and pragmatically take it apart, and even rig up self-made spare parts from improper materials in order to make it work again. This self-assured pragmatism is diametrically opposed to Western images of the impact of technology in Africa and urges us to pay more attention to how machines – not only cars, but also phones, computers, cameras and so on – are used, maintained and repaired in everyday life.[29]

If the car, as many authors suggest (e.g. Mbembe and Roitman 1996: 160), is one of the main markers of modernity, then Kwaku's struggles with *God never fails* can certainly reveal important features of African ways of dealing with modernity. The key term, of course, is 'adjustment' – a process not laid upon from above as in the case of the IMF's Structural Adjustment Programmes, but emerging in the practice of everyday life. 'Adjustment', as we experienced ourselves, becomes necessary as soon as a car enters Ghana and becomes 'part of the system'. This involves ingenious technological changes as well as more or less elaborate spiritual measures. The background of these endeavours, of course, is poverty and scarcity, and it would be naive to merely celebrate them as expressions of cultural creativity. At the same time, at least with regard to the situation described here, there is no reason to subscribe to a pessimistic notion of all-pervasive crisis which speaks through Mbembe and Roitman's account of the Cameroonian situation in general, and their representation of the (middle-class) car as a 'broken down machine' which has been reduced to 'a figurative object' (ibid.: 161) in particular. Kwaku's story reveals a tremendously powerful will and capacity – at least on the part of ordinary people – to keep the engine working by all means, even at a time when the West tends to forget Africa as much as its old, cast-off cars.

We dedicate this chapter to our friend and colleague, the medical sociologist Dr Kodjo Senah, who taught us as much about cars as about Ghanaian society, and with whom we had so many wonderful rides in his Lion (an old Peugeot 505). The research on which this essay is based has been made possible partly through the generous financial support of the *Netherlands Foundation for the Advancement of Tropical Research* (WOTRO). We would like to thank Marleen de Witte for transcribing our interviews with Kwaku and Daniel Miller for valuable editorial comments and suggestions.

References

Agar, M.H. (1986), *Independents Declared. The Dilemmas of Independent Trucking*. Washington, DC: Smithsonian Institution Press.

Date-Bah, E. (1980), 'The Inscriptions on the Vehicles of Ghanaian Commercial Drivers: A Sociological Analysis'. *Journal of Modern African Studies* 18(3): 525–31.

Field, M.J. (1960), *Search for Security. An Ethno-Psychiatric Study of Rural Ghana*. London: Faber & Faber.

Geschiere, P. with C. Fisiy (1994), 'Domesticating Personal Violence: Witchcraft, Courts and Confessions in Cameroon'. *Africa* 64(3): 321–41.

Jordan, J.W. (1978), 'Role Segregation for Fun and Profit. The Daily Behaviour of the West African Lorry Driver'. *Africa* 48(1): 30–46.

Kopytoff, I. (1986), 'The Cultural Biography of Things: Commoditization as Process'. In: A. Appadurai (ed.), *The Social Life of Things. Commodities in Cultural Perspective*. Cambridge: Cambridge University Press, pp. 64–91.

Kyei, K.G and Schreckenbach, H. (1975), *No Time To Die*. Accra: Catholic Press.

Lawuyi, O.B. (1988), 'The World of the Yoruba Taxi Driver. An Interpretative Approach to Vehicle Slogans'. *Africa* 58(1): 1–12.

Lupton, D. (1999), 'Monsters in Metal Cocoons: "Road Rage" and Cyborg Bodies'. *Body & Society* 5(1): 57–73.

Mbembe, A. and Roitman, J. (1996), 'Figures of the Subject in Times of Crisis'. In P. Yaeger (ed.), *The Geography of Identity*. Ann Arbor: University of Michigan Press, pp.153–86.

Meyer, B. (1995) '"Delivered from the Powers of Darkness." Confessions about Satanic Riches in Christian Ghana'. *Africa* 65: 236–55.

—— (1998), '"Make a complete break with the past." Memory and Postcolonial Modernity in Ghanaian Pentecostalist Discourse'. *Journal of Religion in Africa* XXVII(3): 316–49.

—— (1999a), 'Popular Ghanaian Cinema and "African Heritage"'. *Africa Today* 46(2): 93–114.

—— (1999b), *Translating the Devil. Religion and Modernity among the Ewe in Ghana*. IAL-Series. Edinburgh: Edinburgh University Press.

—— (1999c), 'Commodities and the Power of Prayer. Pentecostalist Attitudes Towards Consumption in Contemporary Ghana'. In: B. Meyer and P. Geschiere (eds), *Globalization and Identity. Dialectics of Flow and Closure*. Oxford: Blackwell, pp. 151–76.

Miller, D. (1987), *Material Culture and Mass Consumption*. Oxford: Basil Blackwell.

—— (ed.) (1995), *Worlds Apart. Modernity through the Prism of the Local*. London and New York: Routledge.

Morgan, W.B. and J.C. Pugh (1969), *West Africa*. London: Methuen.

Otnes, P. (1994), *Can we Support Ourselves by Driving to Each Other? Collective and Private Transportation: How the Automobile has Affected Us, What Collective Transportation Does Differently, and Why. A Chreseological Suite in Eight Movements*. Oslo: Institut for sosiologi.

Peace, A. (1988), 'The Politics of Transporting'. *Africa* 58(1): 14–28.

Stoller, P. (1989), *The Taste of Ethnographic Things. The Senses in Anthropology*. Philadelphia: University of Pennsylvania Press.

Van Binsbergen, W. (1999), 'ICT and Intercultural Philosophy. An African Exploration'. Paper Presented to the Seminar 'Globalization and the Construction of Communal Identities', Leiden, February 2000.

Van de Braak, H. (1991), *Een wild dier. De auto als lustobject*. Amsterdam: Uitgeverij Balans.

Van der Geest, S. (1989), '"Sunny boy": chauffeurs, auto's en Highlife in Ghana'. *Amsterdams Sociologisch Tijdschrift* 16(1): 20–39.

Van Dijk, M.P. (1980), 'De informele sector van Ouagadougou en Dakar. Ontwikkelingsmogelijkheden van kleine bedrijven in twee West-afrikaanse hoofdsteden'. (Ph. D. Thesis. Amsterdam: Free University).

Verrips, J. (1990a), 'On the Nomenclature of Dutch Inland River Craft'. *MAST* 3(1): 106–19.

—— (1990b), *Als het tij verloopt . . . Over binnenschippers en hun bonden 1898–1975.* Amsterdam: Het Spinhuis.

—— (1994), 'The Thing Didn't "Do" What I Wanted'. In J. Verrips (ed.), *Transactions. Essays in Honor of Jeremy Boissevain.* Amsterdam: Het Spinhuis, pp. 35–53.

White, L. (1993), 'Cars out of Place: Vampires, Technology, and Labour in East and Central Africa'. *Representations* 43 (Summer): 27–50.

Notes

1. In September 1996, 1000 Cedis were the equivalent of Hfl 1,00, and about US $0.50. Due to inflation, the Cedi has been loosing value to such an extent that in September 1999, C 3400 were exchanged for $1.00. – When we expressed the wish to buy a car, our Ghanaian friends strongly advised us not to get a car which had been 'in the system' for some time already, but rather a second-hand one directly imported from Europe. Despite their sometimes considerable age, such cars are regarded as 'new'.

2. We use pseudonyms for all persons featuring in this chapter.

3. According to Kwaku he painted the slogan *God Never Fails* on the Peugeot, because God had listened to his fervent prayers to influence the previous owner in such a way that he would give or sell his car, which he had stored away for some time, to him.

4. Though Van der Geest even quotes Kopytoff's valuable call in a footnote, he does not take up the gauntlet, for he also deals with cars in a rather superficial way and does not show how they undergo what he tersely calls a 'cultural redefinition' (Van der Geest 1989: 35).

5. Sociologists, too, only paid scanty attention to cars. For notable exceptions see Lupton (1999), Otnes (1994), Van de Braak (1991).

6. Cf. Van Binsbergen (1999), who develops this argument with regard to ICT, and pleads for more research on Africans' use and perception of computers.

7. The only railway which can be said to play a role of some significance is the one between Accra and Kumasi. But in comparison with transport by lorry and car, railway traffic does not mean very much. Though Ghana disposes

over navigable rivers and lakes, especially Lake Volta, waterways are not optimally used for the transport of goods and persons, on the contrary. As a citizen of a country in which transport over water is tremendously important, and as an anthropologist who once did fieldwork among Dutch bargees (Verrips 1990a and b), Jojada often wondered why inland navigation is so under-developed in Ghana.

8. The Ghanaian videofilm industry emerged in the mid-1980s. Instigated by the fact that the state-owned film industry had not been able to produce feature films for years, local, untrained producers took up ordinary video-cameras and made their own films. These movies, which are closer to the soap-genre than to artistic 'African film', visualize stories and experiences from everyday life. They became tremendously popular and are screened in all the major cinemas. Cf. Meyer 1999a.

9. Cars liked very much in Ghana as well as in other African countries are Peugeots. Nowadays they do not reach Ghana over land any longer. Due to the political situation in countries such as Algeria it is too dangerous to drive used cars through the Sahara with the goal to sell them in West African countries for a much higher price than the one paid in Europe.

10. This discouraging policy of the government seems to be successful, for the number of really worn-out and dangerous cars and lorries in Accra's streets in the early 1990s was considerably higher than in the mid-1990s.

11. We also collected in and around Accra a great number of (English, Ga, Twi and Ewe) slogans on cars used for public transport. Hand-painted ones are increasingly being replaced by slogans made of adhesive letters and stickers. Many refer to a Christian background of the driver and/or owner. Islamic ones, such as *Insha Allahu* and *Allahu Wahidu*, are rather rare.

12. At the long-distance stations one often finds a ticketseller, porters, and an overseer or stationmaster who regulates which car will get the next ride and makes sure that the drivers stick to the rules and fares. This system of turns is rather complex and may cause serious conflict between the drivers of 'home' and 'away' cars. Another source of conflict, this time with the passengers, are the extraordinarily high fares drivers sometimes ask on busy Friday evenings for a ride from Accra to other places. The newspapers every now and then publish complaints about this type of 'Friday-evening travel roguery'.

13. In 1996, drivers using an ordinary cab had to pay the owner C 15.000 on each day of the month. In 2000, the amount is C 30.000 (the increase reflects both inflation and higher prices for gasoline). Each month, drivers receive back two-days' pay, and are allowed to keep the extra money made. If the owner lives abroad, the money often has to be paid to a relative, who puts the money in a particular bank account. Car repairs made necessary through old age or accidents form a constant source of dispute between drivers, owners,

and their representatives. The popular soap-like TV series *Taxi-driver* (produced by *Village Communications*, and broadcast by GBC in 1999) depicts this type of conflict. In a comic vein, the series figures the daily experiences of TT, a Ga taxi-driver, with his – sometimes rather weird – customers, the arrogant Asante car owner, car mechanics, policemen, and Christian prophets.

14. In this respect little has changed since Field's observation, made forty years ago, that '[a]mong young men, particularly illiterates, there is no more widespread ambition, nor one more often achieved, than to drive, and if possible own, one of the thousands of passenger car lorries that raven about the roads' (1960: 134).

15. Teshie was originally a Ga fishing village in the immediate neighbour-hood of Accra. In the course of the twentieth century, Teshie grew in size and accommodated an ever-increasing number of persons from other Ghanaian peoples (above all Asante, Ewe, and Muslims from the North) working in Accra. While the old village centre near the coast is still inhabited by Ga fishermen and their families, the newer parts are multi-ethnic. The area where we stayed, and where Kwaku also lives, consisted of a mixture of (run-down) middle-class homes, and compound houses, in which separate rooms were rented to different tenants.

16. With cars as in many professions, the apprentice system is still in operation in Ghana. When a boy wants to become car mechanic, electrician, welder, sprayer, vulcanizer or rubber spare part cutter, to name just a few specializations, he has to become an apprentice in the workshop of a master. In exchange for money and *Akpeteshie* (locally distilled gin) he will usually be trained for four years. Only after having been an apprentice, chief apprentice and assistant, can a young man become master himself. All this time he merely gets 'chop money' and has to serve his master. Officially it is strictly forbidden to offer one's services for money to others without the master's consent, but in practice this happens frequently.

17. This is no small wonder, for the contract contained a clause in which it was stipulated that the car would be the property of the doctor again, if Kwaku would only once forget an installment.

18. A trip with seven passengers from Accra to Takoradi or v.v. brought Kwaku plusminus C 15.000 netto. Together they paid him C 38.500 (C 5.550 each), but he had to deduct from this C 18.000 for petrol, C 1.500 booking fee, C 1.000 welfare fee and C 1.000 union tax. Because gas was cheaper than petrol, Kwaku preferred to use this type of fuel, but often he could not because of problems with the gas installation in his car.

19. The more plates the higher the price. In 1996 the prize of repairing, for example, a 12-volt battery with 19 plates was C 17.500 and refilling plus recharging C 11.500. As in almost all small workshops in Ghana, in battery

repairshops environmental pollution is at the order of the day: acids are just poured on the ground.

20. In this connection it is interesting to take notice of the results of an official German test of vehicles in Kenya (cf. *Süddeutsche Zeitung Magazin* 11.12.98), which showed they all suffered from at least fifty technical shortages. According to the official Kenyan documents, however, some of them were technically OK. We realize that, of course, an examination of cars according to Western standards may be problematic because it fails to acknowledge local technical ingenuity. Certainly there is a great number of 'adjustments' which have no negative impact on safety, although they may still be unacceptable for Western technicians. At the same time, it has to be acknowledged that the line between 'adjustments' and lack of maintenance as a result of poverty may be thin at times. In our view, there are no grounds for an exaggerated cultural relativism, because lives are at stake.

21. Strikingly the introduction of the car did not entail the development of an indigenous Ga, Twi, or Ewe terminology for its parts and their working. Almost all the terms used by drivers and mechanics are English ones.

22. Cf. Stoller who remarks in a chapter on a Songhay (Niger) bush taxi: 'Each taxi, of course has a driver, someone who has obtained his driver's permit and who knows very well how to repair automobile engines' (1989: 73). It is striking, however, that although his anthropology emphasizes the relevance of the senses, he does not say anything about the ways in which such a vehicle is perceived and treated by drivers. Kwaku constantly referred to how he used his senses (especially his ears and nose) to find out how his car was functioning. Striking in this connection were his expressions: 'I felt a smell' or 'I heard a scent'.

23. See Van Dijk (1980:228 ff) who made similar observations for repairshops in Ouagadougou.

24. We found that, in a sense, this field was much easier to grasp than the technological dimension. This does not only stem from the fact that here we touch on common grounds for anthropologists, but also from our own everyday experiences with the representation of and meaning attached to cars in Western societies. Certainly many car advertisements in our own society speak to potential buyers' desire to identify, or better still, be identified with his or her car.

25. In general the number of accidents on Ghana's roads tends to disquietingly increase just before and after Christmas, when many people travel in often overloaded vehicles. It is also in this period that one hears more about molesting and even lynching of drivers who caused serious accidents.

26. We did not hear of rituals performed by groups of drivers in order to protect them while being on the road. In Nigeria, however, cabdrivers are used

to collectively bring blood sacrifices to Ogun, the god of steel whom they consider to be their patron.

27. Passengers and other drivers were not the only possible agents of witchcraft: His wife also feared that beautiful Fanti women on one of his trips to Takoradi would bewitch him so that he would not return to her.

28. Kwaku told us that it was not unusual to purify polluted cars with seawater at the beach because it was supposed to be very powerful. He, however, would not go there, for the seaside was also associated with the Devil and Mami Water (cf. Meyer 1999c).

29. Who hasn't heard the familiar stories about the tractors brought to Africa as a means to further development, which just break down and stop functioning because of a lack of spare parts and know-how? It would be worth to examine as to whether the dysfunctioning may be due to other than technological factors.

Soundscapes of the Car: A Critical Study of Automobile Habitation

Michael Bull

When I get in my car and I turn on my radio. I'm at home. I haven't got a journey to make before I get home. I'm already home. I shut my door, turn on my radio and I'm home. I wind the window down so I can hear what's going on and sometimes as the sun's setting and I'm in town and I think. Wow. What a beautiful city that I'm living in, but it's always at the same time when that certain track comes on. It's a boost. (Automobile user).

Today the highway might well be the site of radio's most captive audience, its most attentive audience. The car is likely to be your most intensive radio listening experience, perhaps even your most intensive media experience altogether. Usually radio is a background medium, but in the car it becomes all-pervasive, all-consuming . . . the car radio envelops you in its own space, providing an infinite soundtrack for the external landscape that scrapes the windshield. The sound of the radio fills up the car, encapsulates you in walls made of words (Loktev 1993).

For twenty-five centuries Western knowledge has tried to look upon the world. It has failed to understand that the world is not for beholding. It is for hearing . . . Now we must learn to judge a society by its noise (Attali 1985).

'. . . to each his own bubble, that is the law today' (Baudrillard 1993).

This chapter discusses the manner in which we lay claim to the spaces that we inhabit through the automobile. While it is commonplace to comment upon the automobile as a technological extension of the driver, in this chapter I focus specifically upon the auditory nature of that technological form of habitation. I do so in order to discuss the specific relational qualities attached to driving to mediated sound in the form of radio or cassette sound. The nature of this aural technological habitation will be discussed through the use of qualitative interview material filtered through an eclectic mix of critical theory.

The metaphor of the car as a home has a long anecdotal history in cultural theory. The root of this is discerned in the automobile as metaphor for the dominant western cultural values of individualism and private property which is coupled to the romantic imagery embodied in travel as signifying individual freedom. The cultural meaning of the automobile as a privatized entity is inscribed into its very origin. From the move away from travelling collectively in trains at the beginnings of the twentieth century (Sachs 1992), to the discomfort of inhabiting restricted spaces with strangers (Simmel 1997) to the desire for smooth, unbroken journeys unfettered by timetables (Urry 1999); these concerns have become embodied in everyday attitudes towards automobile use.[1] The image of mobile and free dwelling on the road denotes the representation of the heroic period of the automobile made mass and democratic. Yet this very democratization of autonomy and control is mediated by the power of the car, hence is dependent. This observation prompted Adorno to remark in 1942: 'And which driver is not tempted, merely by the power of the engine, to wipe out the vermin of the street, pedestrians, children and cyclists?' (Adorno 1974: 40). This description of 'road rage' also signifies the contradictory nature of the automobile embodied in everyday use whereby the driver is simultaneously all-powerful and controlled, not just by the technology of the automobile, but by other drivers and the road system itself. The values embodied in car use also rest uneasily with a growing antipathy to the mundane nature of many aspects of urban everyday life for car users (Sennett, 1990, Bauman 1993). Jacobson, catches the ambivalence of control while at the driving wheel:

> In a car you are physically cocooned . . . It is the last private space in an overwhelmingly public world, the nearest we get to a lavatory at the bottom of the garden, where people once went to have a little time to themselves. Thus does the motor car perpetuate itself. The worse it gets on the roads, the more we seek the solace of our vehicle. This is the way

the world will end – trillions of us fleeing Armageddon, one per car. (Jacobson 2000).

The contradiction appears to be resolved by the nature of automobile habitation itself. Jacobson's metaphor of the garden lavatory is instructive: the car as a private, traditional immobile space of habitation. Yet traditionally Jacobson's escapees retreated into places of silence, to be alone with one's thoughts. Car habitation, in contrast to this, is infused with multiple sounds. The aural privacy of the automobile is gained precisely through the exorcizing of the random sounds of the environment by the mediated sounds of the cassette or radio.

> You have to get it above the noise of the car . . . You switch off to the noise of the car because its the same noise you get all the time. (Sharon)

The automobile as sound environment has been commented upon by Stockfeld:

> The car is one of the most powerful listening environments today, as one of the few places where you can listen to whatever you like, as loud as you like, without being concerned about disturbing others, and even singing along at the top of your voice – the car is the most ubiquitous concert hall and the 'bathroom' of our time.' (Stockfeld 1994: 33).

We can see from this that descriptions of automobile habitation are awash with domestic metaphors of habitation. More recently the car as 'home' is described as becoming increasingly filled with gadgets to make the driver feel even more at 'home'. Automobiles are increasingly represented as safe technological zones protecting the driver from the road and, paradoxically, from themselves.

> I want to suggest that the nature of this 'dwellingness' has changed from 'dwelling on the road' to 'dwelling in the car' . . . Car drivers control the social mix in their car just like homeowners control those visiting their home. The car has become a 'home from home', a place to perform business, romance, family, friendship, crime and so on . . . The car driver is surrounded by control systems that allow a simulation of the domestic environment, a home from home moving flexibly and riskily through strange environments (Urry 1999: 16–17).

There has indeed been a shift in the nature and meaning of automobile habitation which I locate in the development of the automobile as a creator of a mediated soundscape (Schafer 1977). I suggest that the historical turning point between 'dwelling on the road' and 'dwelling in the car' can be located in a very specific technological development: the placing of a radio within the automobile. This has radically transformed the nature of driving and the driver's experience of space, time and place. Thus, an understanding of automobile habitation should focus upon the mediated aural nature of that experience. The custom fitting of radios into cars began in the early 1960s, and by the 1970s were standard equipment in most cars. This was followed by the development of portable tape decks installed in automobiles:

> Tape decks made music consumers mobile, indeed automobile . . . Thus the American mass market was opened up by the car playback system. The mediated sound of the radio and later the cassette deck produce their own specific aural relational qualities (Kittler 1999: 108).

Automobile habitation is usefully understood as representing wider social transformations in which the intimate nature of an industrialized soundworld in the form of radio sounds, recorded music, and television increasingly represent large parts of a privatized everyday lifeworld of urban citizens. This impacts upon habitual everyday notions of what it might mean to 'inhabit' certain spaces such as, the automobile, the street, the shopping arcade or indeed the living room.

> Music was no longer a necessarily public, communal experience, but could be heard at home, divorced from the settings in which it was originally produced. Sound recording, then, gave a powerful boost to the 'privatization' of experience which may have held to be a fundamental aspect of twentieth-century culture (Crissell 1986: 26).

The nature of these everyday aural privatized experiences have, however, remained unexplored, or rather 'invisible' in accounts of everyday life that rather prioritize the visual nature of that experience. Cultural and urban accounts of everyday experience tend to locate experience in a form of visual silence thereby ignoring the specific relational qualities attached to aural experience. These relational qualities will be sketched out in the following section before I describe the everyday nature of auditized automobile habitation.

Sounding out Experience

If the world is for hearing, as Attali suggests, then there exists an unexplored gulf between the world according to sound and the world according to sight. Sound has its own distinctive relational qualities; as Berkeley observed, 'sounds are as near to us as our thoughts' (quoted in Rée, 1999: 36). Sound is essentially non-spatial in character, or rather sound engulfs the spatial, thus making the relation between subject and object problematic. Sound inhabits the subject just as the subject might be said to inhabit sound, whereas vision, in contrast to sound, represents distance, the singular, the objectifying (Jay 1993). Therefore aural relational experience might well differ from a more visually orientated one. This is not to suggest that they are mutually exclusive but merely to suggest that the relational nature of technologized auditory experience differs epistemologically from an explanation that prioritizes the visual. In the following pages I argue that technologies of sound affect automobile drivers' relation to the spaces they inhabit in very specific ways. In doing so I draw upon Lefebvre's understanding of social space together with the early work of Adorno, whose work on technology and the historical construction of the senses has been largely neglected or misunderstood.

What then are the relational qualities of sound in the car and what do forms of aural habitation consist of? In the following pages I draw upon empirical accounts of automobile habitation and reflect on the specifically aural qualities of that experience. The inhabiting of automobile space has cognitive, aesthetic and moral significances that are all relational in so much as they inform us of the ways in which car users relate to their surroundings, others and themselves.[2] By focusing upon the auditory and the technologized nature of everyday experience of automobile users, I explain their attempts at creating manageable sites of habitation.

Sounding out Automobile Habitation

The proximity of the aural defines car habitation for many drivers. Drivers often describe the discomfort of spending time in their cars with only the sound of the engine to accompany them. Driving without the mediation of music or the voice qualitatively changes the experience of driving. Many drivers habitually switch on their radio as they enter their automobile describing the space of the car as becoming energised as soon as the radio or music system is switched on:

In the mornings I feel relaxed when I get into my car. After rushing around getting ready, it's nice to unwind, put on my music and the heater and get myself ready for the day. (Jonathan)

I suppose I feel at ease. I put the radio on, put the keys in the ignition and I'm away. I've had new furry covers put on the car seats, so they are really comfortable and snug. In a way too, I suppose after getting out of the house, getting into Ruby (the car) is a way for me to relax and unwind. (Alexandra)

It comes on automatically when I switch the ignition on. Like I never switch the power off, so it automatically comes on as soon as I start the car. (Alicia)

Well it's on anyway. When the car starts it switches on. So it comes on automatically. (Gale)

I can't even start my car without music being on. It's automatic. Straight away, amplifiers turned on. boom boom! (Kerry)

Mediated sound becomes a component part of what it is to drive. The sound of music competes with the sounds of the engine and the spaces outside of the automobile. Adorno in his early work describes this aural proximity in terms of states of 'we-ness' which refers to the substitution of 'direct' experience by a mediated, technologized form of aural experience. Music or recorded sound is both colonizing and utopian, according to Adorno, as it increasingly fills the spaces of 'habitation' in everyday Western culture. Recorded sound, according to Adorno,

takes the place of the utopia it promises. By circling people, by enveloping them – as inherent in the acoustical phenomenon – and turning them as listeners into participants, it contributes ideologically to the integration which modern society never tires of achieving in reality. It leaves no room for conceptual reflection between itself and the subject, and so it creates an illusion of immediacy in the totally mediated world, a proximity between strangers, of warmth for those who come to feel the chill of the unmitigated struggle of all against all. Most important among the functions of consumed music – which keeps evoking memories of a language of immediacy – may be that it eases men's suffering under the universal mediations, as if one were still living face to face in spite of it all. (Adorno 1976: 46).

Without necessarily endorsing the implied totalitarian nature of technology in the above quote, Adorno's insight is significant in its highlighting of the auratic quality of music together with its integrative and utopian function. The subjective desire to transcend the everyday through music becomes a focal point of his analysis, as is the desire to remain 'connected' to specific cultural products. The nature of this 'connection' constitutes the state of 'we-ness'. The 'social' thus undergoes a transformation through the colonization of representational space by forms of communication technology, and the 'site' of experience is subsequently transformed.[3] Increasingly for Critical Theorists the technologically produced products of the culture industry, in all of its forms, become a substitute for the subject's sense of the social, community or sense of place. This, for Adorno, produces consumers who become increasingly 'addicted' to using those products which act as a substitute for the above. Central to this is a transformed notion of relational experience:

> Addicted conduct generally has a social component: it is one possible reaction to the atomisation which, as sociologists have noticed, parallels the compression of the social network. The addict manages to cope with the situation of social pressure, as well as that of his loneliness, by dressing it up, so to speak, as a reality of his own being; he turns the formula 'Leave me alone!' into something like an illusory private realm, where he thinks he can be himself (Adorno 1976: 15).

Music, for Adorno, increasingly fills the gap left by the absence of any meaningful sense of the experienced social. Technology is perceived as paradoxically enhancing and increasingly constituting that impoverishment which, for Adorno, contributes to the dependency of the user/ listener. Music as such becomes a substitute for community, warmth and social contact. In this isolated world of the listener a need arises to substitute or replace his or her sense of insecurity with the products of the culture industry, leading to new forms of dependency. However it is specifically from Adorno's recognition of an 'unfulfilled' articulated through the auditory that a potentially active formation of agency and intentionality might be developed. The aural nature of 'we-ness' can be charted in the behaviour of automobile users in their everyday experience. For many users the radio/sound are integral components of automobile habitation:

Well, the stereo's the most important thing in my car. I don't like driving especially if I'm doing long distances on the motorway. I have to have music. (Alicia)

It's lonely in the car. I like to have music. (Joan)

I don't like the dark anyway. But when it's really late at night and the streets are really deserted, I think that does help. You know. You turn it up or whatever and at night when I have the radio on, they have like 'Late Night Love' and it's just hearing other people's voices, and it's actually quite funny listening to some of the stuff they say. I think if I didn't have that I would be a bit more freaked out about driving late at night. (Alice)

I put it on to Radio 4, because I knew I had a long drive. So it depends what kind of drive, Radio 4, I wanted someone talking to me, yes, I need someone talking to me, so I put it on Radio 4. I want to be listening to a voice telling me about various bits of news, etc. (Sharon)

The talking stations are very much to key yourself into the world – to engage (John)

It connects me to the world because you've got someone talking to you, to connect you. (Ben)

Automobile users thus appear to prefer inhabiting an accompanied soundscape. However, for many drivers, this auditized space is preferably one occupied solely by themselves. Drivers often mention that the space of the car is preferably a privatized space.

As a private sphere the car is first and foremost a transportation environment, different in kind from the living, leisure and working spheres. For many drivers the time spent in the car may be the only regular opportunity for reflection, voluntary solitude and concentrated listening to the radio or to recorded music, the only chance to do nothing without having to appear to be doing something else. In the car you can be at peace, be whoever you like. (Stockfelt 1994: 30).

It's a totally different environment. I like driving, I love driving on the road and you like driving on your own because it's a totally separate environment. It's a total indulgence, it's your environment. You control it. You can do whatever you like in it. It's like time off. You're travelling from A to B, but it's the ultimate in idleness, really. You're not really doing anything but listening. It's great . . . So when there's someone else in the car with you you don't have so much control over the environment. You can't let go so much. (Sharon)

The privatized aural space of the car becomes a space whereby drivers reclaim time, away from the restrictions of the day. The mundane activity of the day is transformed into a personally possessed time. Listening to music/radio enhances the drivers' sense of time control/occupancy.

I always have the radio on when I'm driving on my own and I'm always really annoyed when people in the car with me don't want the radio on. I just really enjoy, especially if I'm doing a drive that I do regularly, that, I can sort of switch off and also if I'm doing it at a similar time. I can key into the Radio 4 schedule, and I know where I am in reference to whatever's on the radio. (Susan)

It's my time. I'm at home. I feel that I'm in my own space. More so than I would be in a tube listening with a Walkman. I feel totally at home. I could stop the car in the middle of a lane and eat in my car . . . A lot of people used to say: Joyce doesn't have a room. She has a car. (Jo)

It's much more convenient to use my own car. I can then relax, listen to the Breakfast Show on Radio 1 and get into my own time. (Alexandra)

The sound of the radio voice or music fills up or overlays the contingency of driving, transforming the potential frustration associated with powerlessness into pleasurable, possessed time:

It's good if you're in a traffic jam cos you can just forget where you are and just listen to it. It's more really background though. I usually sing along. (Gale)

I couldn't drive without music. Driving without music is too boring. (John)

Sometimes you can just lose an hour, I can completely forget say between Swindon and Cheltenham. It will be like, did I do that? Because I've been so involved in my music. It's like I'm on autopilot. (Lizzy)

It's really boring in a car without sound. I don't want to go on a journey in a car without a radio. (Mark)

Not only does time pass more pleasurably, it is also potentially more predictable:

I always go at a certain time, so I always listen to the same programme, a dance programme. I look forward to it. The relax programme! (Sharon)

It's important that the same programmes come on at the same time. I feel I start the day when I listen to the day's Story at 9.45. (Susan)

I have a few favourite sing along tapes. These are for long journeys. I will have them on the side, on the passenger seat, a selection of tapes because you can't rely on the radio for such a long journey. (Lizzy)

The structure, duration and meaning of the journey is increasingly mediated by sound. While automobile drivers speak of the car in terms of 'home' the aural space of the car differs from that of the living room. Radio programmes or music listened to in the car do not repeat patterns of domestic consumption.

I think there's something specific about the car. There's a certain kind of sound you can achieve in the car that you can't at home. I think it's the smaller space. I just don't get the same feeling at home as I do in the car. I don't have the music on as loud at home. (In the car) I feel freer. (James)

I play it louder if I'm on my own and I let myself go a bit more if I'm on my own. (Gale)

Music or sound is chosen to suit the time of day or the mood of the driver.

I have to have it on loud [Radio 1] Each track identity free, repetitious, high energy, that makes me drive very carelessly! (Susan)

I'm more likely to listen to music after work on the way home or when I'm doing something else because its much more relaxing and it's a different type of music as well. I'll listen to a dance station, such as Kiss or Radio 1 if they've got some good dance music on and that will make me feel good. And then there's tapes. Tapes are reserved for those times when there's nothing on the radio – Then I've got my selection of tapes, and for long journeys – tapes are a way of recording how far you've got. So you change every 45 minutes. When you're on your own – so you say, That journey will take four tapes. (Sharon)

Many drivers take a selection of tapes so as to be able to match their mood to music. The music both enhances the time of the journey – the experience of driving itself – and enables the driver to control his or her thoughts:

It's about switching off, there's also the pleasure if I know I'm going out. Music is a way of shaking things off. (Susan)

The pleasure is definitely in the listening. I sometimes go and slap it on and get away from reality. (James)

The success of music/sound in transforming the experience of driving is largely dependent upon the intensity of sound. This means that all other sounds have to be masked by the aural sounds in the car:

If I'm on my own then I do have it on pretty loud. (Jo)

Especially along a motorway, I'll have it really loud, cos there's nothing to do, I'm usually on a motorway when its dark anyway so I like to. It's really nice if I'm on my own it's nice company, so I have it up loud. (Kate)

The aural space of the automobile thus becomes a safe and intimate environment inhabited by the mediated presence of consumer culture. Making the mobile, contingent nature of the journey into one which has the appearance of precisely the opposite of this. However this aural form of habitation often includes interaction with the drivers themselves contributing to the soundscape of the automobile.

Singing in Car

The car is a space of performance and communication where drivers report being in dialogue with the radio or singing in one's auditized/ privatized space. Baudrillard's bubble (Baudrillard 1993) is however a fragile one, in which even aural absorption doesn't fully protect the aural bubble of habitation. The space of a car is both one to look out from and to be looked in to. It is simultaneously private and public. Drivers both lose themselves in the pleasure of habitation and become increasingly aware of the 'look' of others. Many drivers sing or talk to their radio while driving:

> Actually that's one thing I love about my car – she's all mine. I don't have to share her with anyone. I can do what I like in my car – with reason – I can turn the radio up full blast and have a good sing song without anyone looking at me. Actually, sometimes I suddenly realize that I'm merrily singing along, and the person in the next car is having a good laugh at me, but I forgot that people can see in and I get really embarrassed. (Alexandra)

> I'll sing along at the top of my voice and I always worry what people in other cars think when they see me. They think I'm talking to myself or something . . . I just sing along all the time. I don't stop, like every song that comes on. Cos I watch a lot of music channels at home, so I know the words to a lot of songs. If I'm listening to the radio, I'll sing along to practically every song that comes on. (Alicia)

> It's the embarrassment factor. When you're in your car and you get caught talking at the radio, dancing or smiling – it happens so regularly. When you're on the motorway, there's no one there, but in town, it's different. In town you switch into the fact that you're visible. (John)

> I'm sitting there mouthing off to it. You talk, as you would any time when you're on your own. If the TV's on and there's some news programme, I'd talk, like that's a load of rubbish, etc., I'd chatter away to it. (Sharon)

The space of the car becomes a free space in which the driver feels free to indulge his or her aural whims with no inhibitions. Houses have other occupants or neighbours to inhibit any such desire:

> The louder the better. In fact, I use my car, I use it more than in the house, because I don't want to annoy the neighbours. But in the car, traffic is very noisy, so nobody can hear you. I sing incredibly loudly, especially on the motorway – In fact I have certain cassettes that I'll put on to sing incredibly loudly to' (Susan)

The sound of music together with the sound of their own voice acts so as to provide a greater sense of presence as well as transforming the time of driving. Drivers even in the act of singing are not of course hermetically sealed from the outside world. The very act of driving requires a steady focus upon the road. Outside space becomes the space of the road before the driver. Drivers may well 'look' to music but more often report the intimacy of sound in their space. Sometimes this auditory home encompasses the aestheticization of the world outside of the automobile:

> When I'm sat in a traffic jam or at traffic lights, in town especially, to ease the boredom, I quite enjoy watching what's going on around me. I look in other people's cars, and watch people walking down the street. I like to see what they're doing and where they're going. As I am in my car a lot, I do need something to take away the boredom. The radio is good for that too. Actually I find music in the car changes how I look at the outside. It entertains me to watch other people with my music on. It is as if they are walking along to the music. (Richard)

Sound thus acts to transform the representational space of habitation, both within and outside of the automobile.

> You and the car are one thing and that's it and that's your space. Outside its different. You're in your time capsule, it's like your living room, your mobile living room. (Sharon)

> It's an extension of my space when I'm on the road. (Susan)

Drivers do experience their space as one which is hermetically sealed:

> When you're in your car you don't notice the pollution even though you're in your car which is polluting the atmosphere. Somehow you don't notice it. And I do find when I'm in London and I'm on foot and I'm having to walk and get a tube I'm much more tired at the end of the day and actually have a headache. It does get to me a little bit more. The chaos of the city. Whereas in my car I can be stuck in a traffic jam for three hours and it doesn't affect me. I stay calm. (Jo)

> I think you're in your own little bubble. You're in your own little world and you have a certain amount of control and you don't have so much interruption. (John)

> When I'm in my car with the radio on, nothing outside seems to matter. It's like I'm the only one who is really there, and everyone else, the drivers,

the people walking by are not kind of real. I suppose it seems like that because I am shut off from other drivers. They don't seem real. (Alex)

This hermetically and aurally sealed form of living in a privatized public space also affects the drivers relation to the act of driving. Driving is invariably described as more pleasurable when accompanied by music or radio sound. While some drivers report avoiding certain kinds of music as they feel they might drive too quickly or get 'carried away' by the sweeping or emotional force of the music, this is by no means universal. Equally drivers sometimes report moving and manoeuvring through traffic in a dance-like manner, as if the relation between the driver and the act of driving were essentially aesthetic. Descriptions of driving often take on a romantic or filmic resonance in the literature:

Who can resist keeping the station tuned to 'Born to be Wild' whilst racing down the interstate. Crankin' it up. Firin' up a cigarette. Rollin' down the windows. Exceedin' the speed limit . . . Dreamin of automotive decadence. (Locktev 1993: 206).

While this is of course one structural possibility of the aural experience of driving, it is incorrect to stereotype the aural experience of driving under this heroic banner. More interestingly the following driver describes the simultaneous nature of listening and driving in which the private experience of listening is seen as paralleling the public occupancy of the space of the road:

It's very strange, what happens. You're driving on a very low level in terms of your awareness of you're driving and you drive on this low level . . . This whole conversation is going on in your head with the radio which is on a totally separate level and you're absolutely, you feel 100 per cent aware of both. You're quite capable of taking in information from both, but they're both separate. (Sharon)

The Aural Management of Space and Time in the Car

Drivers spend an increasing amount of their daily lives in automobiles, often involved in mundane and repetitive journeys. The above analysis demonstrates a variety of practices in which the daily scripts of experience are continually rewritten. These areas of everyday life, undertaken in public spaces, have often been assumed to be 'unscripted' and void of interest. Automobile habitation rather sees inhabitants rewriting their daily script through the mediation of sound. The aural script of 'driving time' is imposed upon those mundane and routine periods of empty time, thereby reclaiming and transforming them. Alternatively, the 'script' is extended into linear time in order to delay involvement in the 'bad' script of the unpleasurable and inevitable everyday. The need to reappropriate time and script it in as many activities as possible is indicated in the above analysis of automobile use. Automobile habitation permits an analysis of the way in which drivers redraw the meaning of journey times thus confounding traditional dichotomous definitions of the partitioning of the everyday into meaningful/meaningless time:

> We can happily become lost in the anonymous ritualised journeys to and from work, for we know that this surrenders nothing that is important to ourselves. We do not live or do identity work in these places. Real life is elsewhere. (Cohen and Taylor 1976: 50).

Automobile habitation, with its privatized sound world, rather represents a form of 'compensatory metaphysics' in which time is transformed and experience heightened. The aural habitation of place and time becomes a way of reinscribing the ritual of everyday practices with the driver's own chosen, more meaningful set of 'rituals'. What becomes clear is that the notion that 'real life is elsewhere' is experienced negatively by many automobile users who use the radio or the cassette as a means to reclaiming significance in the present.

Automobiles appear to operate as symbolic 'sanctuaries' in which drivers operationalize strategies of 'centredness' (Sennett 1990). This sanctuary represents a physical zone of 'immunity' between the driver and the world or space beyond. Historically this zone was thought to be imbued with qualities not attributable to the world beyond, as in the spaces of a church. Sennett attributes this withdrawal to a recognition that the urban world is one of confusion and instability, lacking in any clear definition. The attempt to create order, stability and control within some 'inner' realm is understood in terms of a progressive 'privatisation of experience' in the work of Sennett. Following Sennett, the automobile might constitute one of the last, albeit problematic, refuges of a retreating public subjectivity. The 'automobile sanctuary' is conceptually enhanced through privatized listening which erects a convincing and intimate barrier between the subject and the exterior world. Automobile users consistently refer to the car through the metaphor of 'home'. Yet a home in which they are preferably the sole occupant accompanied by the sounds of the radio or CD. Adorno frequently dwells on the historical process in terms of the auditory 'colonisation' of the 'site' of experience by the social network:

> We might conceive a series leading from the man who cannot work without the blare of the radio to the one that kills time and paralyses loneliness by filling his ears with the illusion of 'being with' no matter what. (Adorno 1991: 16).

Adorno points to the problematic nature of the 'site' of experience whereby the subject dresses up his 'site' which is actually an 'illusory realm' filled out by society. However, the above analysis indicates that automobile habitation is not merely 'passive' in its constitution. While Adorno's analysis of auditory experience, as does Lefebvre's understanding of representational space, drifts towards notions of the colonization of experience, an empirical understanding of automobile use suggests a more dialectical process in which drivers actively reconstruct their experience precisely through the mediated sounds of the culture industry. Such an analysis sheds light on the ambivalent relation to public spaces embedded in the everyday use of the automobile.

Notes

1. Sachs notes that 'Far from being a mere means of transport, automobiles crystallise life plans and world images, needs and hopes, which in turn stamp the technical contrivance with a cultural meaning. In this interchange, culture and technology prove mutually reinforcing. Technology does not fall from the sky; rather, the aspirations of a society combine with technical possibility to inject a bit of culture into the design like a genetic code. Yet neither do lifestyle and desire emerge from the thin air of culture; instead they coalesce around a given technology. (Sachs 1992: 92).

2. 'Social space ought to be seen as a complex interaction of three interwoven, yet distinct processes – those of cognitive, aesthetic and moral "spacings" and their respective products' (Bauman 1993: 145).

3. If car use transforms the relationship between the driver and urban space then this relationship can be located conceptually within Lefebvre's understanding of representational space, by which he means:

Space as directly lived through its associated images and symbols, and hence the space of 'inhabiters' and 'users' . . . This is the dominated – and hence passively experienced – space which the imagination seeks to change and appropriate. It overlays physical space, making symbolic use of its object . . . Representational space is alive: it speaks. It has an effective kernel or centre: Ego, bed, bedroom, dwelling house; or square, church, graveyard. It embraces the loci of passion, of action and lived situations . . . It may be directional, situational or relational, because it is essentially qualitative, fluid and dynamic (Lefebvre 1991: 39–42).

The 'site' of experience exists for Lefebvre within representational space. This can be described phenomenologically in terms of the direction, situation and relation of the experiencing subject. Lefebvre's analysis is able to accommodate a qualitative, multi-layered and dynamic evaluation of experience in relation to its surroundings. However, in doing so Lefebvre appears to create an either/or dichotomy that sits uneasily with his otherwise fluid analysis of experience.

References

Adorno, T. (1974), *Minima Moralia: Reflections on a Damaged Life*. London: New Left Books.

—— (1976), *Introduction to the Sociology of Music*. New York: Continuum Press.

—— (1991), *The Culture Industry: Selected Essays on Mass Culture*. London: Routledge.

—— (1992), 'Fetish Character in Music and Regressive Listening'. In A. Arato and E. Gebhardt. *The Essential Frankfurt School Reader*. New York: Continuum Press.

—— and Eisler, H. (1994), *Composing for the Films*. London: Athlone.

Arato, A. and Gebhardt, E. (eds) (1992), *The Essential Frankfurt School Reader*. New York: Continuum Press.

Attali, J. (1985), *Noise: The Political Economy of Music*. Minneapolis: University of Minnesota Press.

Bachelard, G. (1994), *The Poetics of Space: The Classical Look at how we Experience Intimate Places*. Boston: Beacon Press

Baudrillard, J. (1993), *Symbolic Exchange and Death*. London: Sage.

Bauman, Z. (1993), *Postmodern Ethics*. Oxford: Blackwell.

Caygill, H. (1998), *Walter Benjamin: The Colour of Experience*. London: Routledge.

Cohen, S. and Taylor, L. (1976), *Escape Attempts: The Theory and Practice of Resistance to Everyday Life*. London: Routledge.

Crissell, A. (1986), *Understanding Radio*. London: Routledge.

Frisby, D. and Featherstone, M. (eds) (1997), *Simmel on Culture*. London: Sage.

Horkheimer, M. and Adorno, T. (1973), *The Dialectic of Enlightenment*. London: Penguin.

Jarviluoma, H. (ed.) (1994), *Soundscapes: Essays on Vroom and Moo*. Tampere: Tampere University Press.

Jay, M. (1993), *Downcast Eyes: The Denigration of Vision in Twentieth-Century French Thought*. Berkeley: University of California Press.

Kahn, D. and Whitehead, G. (eds) (1992), *Wireless Imagination: Sound, Radio and the Avant-Garde*. Cambridge, Mass.: MIT Press.

Kittler, F. (1999), *Gramophone, Film, Typewriter*. Stanford: Stanford University Press.

Kracauer, S. (1995), *The Mass Ornament: Weimar Essays*. Cambridge, Mass: Harvard University Press.

Lasch, S. (1999), *Another Modernity, A Different Rationality*. Oxford: Blackwell.

Lash, S. and Urry, J. (1995), *Economies of Signs and Space*. London: Sage.

Lefebvre, H. (1991), *The Production of Space*. Oxford: Blackwell.

Loktev, J. (1993), 'Static Motion, or the confessions of a Compulsive Radio Driver'. In *Semiotexte* Volume VI, 1.

O'Connell, S. (1998), *The Car in British Society: Class, Gender and Motoring 1896–1939*. Manchester: Manchester University Press.

Rée. J. (1999), *I See A Voice: Language, Deafness and the Senses, A Philosophical Enquiry*. HarperCollins: London.

Sachs, W. (1992), *For Love of the Automobile: Looking Back into the History of our Desires*. Berkeley: University of California Press.

Schafer, R. (1977), *The Tuning of the World*. New York: Knopf.

Sennett, R. (1990), *The Conscience of the Eye*. London: Faber & Faber.

Simmel, G. (1997), 'The Metropolis and Mental Life'. In D. Frisby and M. Featherstone, *Simmel on Culture*. London: Sage.

Stockfeld, O. (1994), 'Cars, Buildings, Soundscapes'. In H. Jarviluoma, *Soundscapes: Essays on Vroom and Moo*. Tampene University Press.

Urry, J. (1999), *Automobility, Car Culture and Weightless Travel*. Lancaster University at http:/www.lancaster.ac.uk/soc030ju.html

Negotiations of Car Use in Everyday Life

Simon Maxwell,

Introduction

Meanings of car use are embedded in social and cultural relations, yet the intricate ways in which these meanings are created and maintained through interactions with others and negotiations with the self in everyday life are often neglected. In contrast, many of the utilitarian benefits of car use are well recognized. In the last few years, anthropological approaches to consumption have helped us to understand how the *social* meanings of commodities, essentially inanimate goods, are far more significant to the ways in which our society works than the *utilitarian* meanings of these goods (Douglas and Isherwood 1996; Fine and Leopold 1993).

Meanings of car use in everyday life are increasingly intertwined with deep public concerns for both social and environmental consequences of increasing levels of car use. In this chapter, I argue that these concerns coexist with positive meanings of car use embedded in social relations which are only articulated in an extremely fragmented form. It is somewhat ironic that existing public policy and government rhetoric in the UK seems to neglect meanings of car use in both these respects.

This chapter explores ways in which people 'make sense' of personal car use in the face of what is sometimes termed our 'car dependence' (Goodwin 1995), and a collective sense of unease about the increasing social and environmental problems associated with car use. It argues that many constructions of meaning around car use seek to reduce guilt and anxiety experienced in relation to increasing levels of car ownership and use.

Many other largely social ethics are also deeply intertwined with car use in everyday life. These include thriftiness, saving time, work, and looking after and caring for immediate others, friends and neighbours. Car use is associated with intense negotiations in everyday life, between these ethics and the social and environmental anxieties outlined above. Finally, the chapter briefly considers the need for new deliberative and inclusionary decision-making processes through which both individual and collective issues and concerns associated with car use may be articulated and addressed.

Overview of Study

This chapter is based on a study using two in-depth discussion groups to explore public meanings of car use. Each in-depth group had ten members and met once a week for five weeks. Each meeting lasted one and a half hours. The groups were conducted according to theory and methodology of psychoanalytic group work, specifically with a relatively unstructured approach, and an emphasis on creating a shared space where people felt able to trust one another and explore issues of interest and concern to them (Burgess *et al.*, 1988a, 1988b; Harrison and Burgess, 1994). One in-depth group was carried out with parents of young children (Group A) and the other with people aged fifty or over (Group B). This strategy aimed to encourage discussion of issues and meanings of car use that are largely excluded from existing decision-making processes among similar groups of people. Following the group meetings, an individual follow-up interview was conducted with each participant to discuss any further issues that he or she wished to raise and general experiences of the group. The groups were conducted in Cambridge, a city where transport issues are a major concern in everyday life and to decision-makers.

Participants were recruited largely through door-to-door recruitment in one area of the city, but also through local groups for parents and young children, and through friends, neighbours and partners of these initial contacts. Door-to-door recruitment was used to ensure as broad a range of people as possible participated in the work, and specifically to ensure that the work reached out beyond the enthusiastic people who are members of many local groups, who are generally well-connected and who dominate many participatory initiatives. It was also considered important to ensure that the work reached out beyond people with a particular environmental or other political stance regarding car use.

The concept of 'discursive repertoires' (Jackson *et al.*, 1999) was used during analysis of material from the in-depth groups. In a recent paper exploring meanings of men's lifestyle magazines, Jackson *et al.* employed a number of 'discursive repertoires' identified from focus group discussions. The authors state that there are parallels with Frazer's (1992, cited in Jackson *et al.*, 1999) concept of 'discourse register': 'an institutionally, situationally specific, culturally familiar, public, way of talking'. Jackson *et al.* state that they, like Frazer, are interested in the repertoires that are available to people, and in the ways in which these structure and shape the things that people say in different contexts.

Four 'discursive repertoires' were developed through analysis of transcripts from the groups. There are considerable ambiguities, contradictions and ambivalences both within and between each of the 'discursive repertoires' outlined below:

- utility – advantages of car use include the ease and convenience of getting around by car for many journeys, as well as the fact that it is often cheaper than other modes of transport at the point of use. This ease and convenience stands in contrast to increasing experiences of congestion, particularly on journeys to and from work and at other peak times;
- freedom and independence – meanings of car use associated with freedom and independence are still very strong, although these meanings are often neglected in discourses of car use outside advertising. These meanings stand in stark contrast not only to direct experiences of congestion, but also to intense feelings of disempowerment over wider transport choices and other political decision-making;
- risk – car use offers reduced personal risk for many people, particularly when travelling at night, but also presents increased individual and collective risks in terms of personal health and through road traffic accidents. The deeply permeated and uncertain environmental risks associated with growing overall levels of car use are also considerable; and
- social meanings and negotiations of car use – the principal focus of this chapter, although material from the first three 'discursive repertoires' is discussed in brief where appropriate.

'Making Sense' of Personal Car Use

People are aware that car use is associated with significant social and environmental problems, and this awareness is tightly interwoven with meanings of car use in everyday life. It is difficult to talk about the freedom and independence associated with car use without immediately adding some qualification about congestion. Similarly, it is difficult to talk about ensuring the safety of children by taking them to school in the car without also considering that overall levels of car use make it increasingly unsafe for other children to walk or cycle to school on their own. Many of these problems are well rehearsed, and are therefore not discussed in any detail in this chapter.

The ways in which participants 'make sense' of car use in the face of these concerns for the social and environmental consequences of their individual and collective action is deeply intertwined with a desire to reduce anxiety and guilt. Warde (1994) argues that a perceived lack of choice over consumption behaviour can mitigate anxiety. Alice, a woman in her thirties with one young child, felt that 'the problem's out of our hands':

> I mean, my mother, she never worked, she took us out, she went and did the shopping everyday and walked back home, we never had a car. My father, you know, did a nine to five job. How many people now do a nine to five job? How many people aren't rushing to the childminder, to work, to home, to Sainsbury's and everything else, and you, sort of, it's a . . . almost a *fait accompli*, you have to use the car to keep up with the pressure of twentieth-, late twentieth-century life (A1, 1710–1717).

Sanjeev, a man in his forties with two young children, spoke about lack of choice over jobs, and about how this can affect transport choices. He had moved from London with his job four years previously, and drove about eight miles out of the city to work:

> I mean, because there's so much competition . . . you can't really choose a job near your house, or within easy reach of the transport, isn't it, these days, if you get a good offer, good job . . . you'll take it up, won't you, so you can't really choose some things, like you said about school, something, yes . . . if you know it's a good school, you want your child to go there, so you will drive and drop the child off, same thing with jobs (A5, 990–997).

Anne, a woman in her fifties, discussed what would happen if she gave up her car:

> The thought that if I or somebody else would give their car up, there would be somebody else who would take my spot or, you know, the

length of [. . .] Road, somebody else would come and park their car in the space, you just think, well, why? (B3, 1786–1790).

Participants employed a range of constructions which seemed primarily intended to reduce personal responsibility by transferring blame to others. Richard, a man in his forties with two young children, drove around nine miles to his work in a village outside Cambridge. He was keen for the social benefits resulting from employment opportunities such as his own to be recognized:

I work in [. . .], in the village of [. . .], and . . . there's sort of converted farm buildings and things, into office areas and stuff, and then further into the village there's another sort of development of units behind the post office and store and, well, [. . .] is a good example of where it could do with a little bit more in terms of people and development because everything's dying. The post office and stores in [. . .] did close down, it has reopened and it's doing all right but, you know, it was very close to being a village with no shop (A1, 863–875).

Through this construction, Richard casts his own commuting in a socially advantageous light. As Richard talked more about his journey to work, it became clear how meanings associated with the social regeneration of the village were also intertwined with his perceptions of others' use of their car. He explained how to him, others use the car in, literally, the 'wrong' way, whereas his journey, against the flow of traffic into Cambridge, is described as a 'spin':

I mean, the car is a wonderfully useful tool, and attachment, but unfortunately it's used for much of the time in the wrong way . . . When I spin out to [. . .] everyday I see the three-mile queue up past the American Cemetery and they're all queuing, one person to a car to get into work (A1, 1769–1773).

Responsibility is also delegated within households. Bill, a man in his forties with three children, said that his partner is 'totally car bound', and in a very open and self-deprecating manner, admitted that he finds this 'quite convenient really'.

Some participants seemed unwilling to accept any responsibility for problems associated with car use, and did not seem to experience guilt or anxiety over their own personal car use. Convoluted constructions of meaning may be required in order to maintain this lack of anxiety.

Joyce, a woman in her fifties, discussed the positive social benefits of taking her car to work:

> I take my car to work so that the people who work at the Cambridge Building Society who all arrive by car, and the hairdresser who drives in from [...], can actually have my space to park in during the day (B3, 1146–1149).

Joyce also shared some of her feelings about the relationship between humans and nature. She received some support from Anne, although Catherine and Liz, also in their fifties, contested the lack of willingness to accept any collective responsibility for the environmental consequences of car use:

> Joyce: [D]on't you think that nature as a whole is much stronger than anything human beings do? You only want one huge volcanic eruption and it throws far more into the air than a lot of your cars . . .
> Catherine: Yes, but that would be added on, it's not instead . . .
> Liz: Yes, they've always been there . . .
> Joyce: But I think you need to keep in proportion, you know, human beings have an enormous, umm, idea of how important they are. We're only passing through . . .
> Anne: Yes, absolutely . . .
> Liz: But it is this, what we're contributing, that is making the difference to the global warming and that . . .
> Joyce: Well, perhaps it's part of evolution . . .
> Liz: But it's our conscious . . . in a sense, we're deliberately changing that. I wouldn't put it down to, just sort of like, by the by, evolution, like, we're animals, we're not aware of it. We are, we know what we're doing, but we're not actually taking the steps to really consciously stop that (B3, 2276–2300).

Many participants had made significant changes in their behaviour in order to limit or reduce their own personal car use. Although the social and environmental benefits of such actions should be acknowledged, they also clearly serve to reduce guilt and anxiety.

Several participants discussed their infrequent use of the car. In a typical statement, Claire, a woman in her thirties with two young children, said that their car 'sits outside the house all week and we use it to go to Sainsbury's, or something like that, but we rarely use it'. Julie, a woman in her twenties with one young child, had made a deliberate decision not to have a car. She acknowledged that it was 'very difficult really' if you had to 'get by' on public transport, but stressed the efforts that she would continue to make to ensure that she did not need a car:

> I really like it now, that I can just get about locally on my bicycle, and I would make a real effort to arrange that my life would be like that, and in the future that I could walk to school and walk to my job, 'cos it gives me sort of far better quality of life than it does zooming round in a car . . . sitting in traffic, just hated it (A4, 1096–1101).

Liz highlighted the importance of the environment to her, alongside her own personal health, in her decisions to find alternatives to the car:

> [M]y decisions are based environmentally as well as personally to do with my health. Now, you know, I've told you I walk quite a lot but towards the beginning of the week my work was taking me quite a lot in the car, and then I had a day where I had an appointment in town, I had an appointment on [. . .], and I had an appointment at [. . .], I had an appointment at [. . .] and I had an appointment at home, and I decided that morning, because I'd been in the car for the last three days, my . . . I was starting to get tingling up and down my arm, and I thought, 'Uh, oh. Stress' [Liz laughs], and then I thought, 'When can I get some exercise in', and the weather forecast wasn't too bad that day, I thought, 'Right', so I walked the whole lot and, you know . . . it was a good amount of miles, I can tell you (B4, 1554–1567).

Four principal themes have emerged to this point through which people are struggling to 'make sense' of personal car use. All the constructions and actions discussed clearly serve to reduce anxiety and guilt experienced in relation to car use. Many participants did not feel they have any real choice in the use of their car. Others did not acknowledge any sense of agency in assessing the consequences of car use. Many participants had found ways in which to delegate responsibility for the social and environmental consequences of car use, often to other car users as well as to government. However, many people had also made great efforts to limit or reduce their car use. For these participants, the social and environmental benefits of finding alternatives to the car intersect with other varied concerns, ranging from personal health to general quality of life.

Meanings of car use in everyday life may also serve subtle social functions. Bourdieu (1984) explores the creation of meaning around patterns of consumption to construct and maintain social difference or 'distinction'. Miller (1987) discusses the ways in which material objects may represent and assist in the construction of perspectives relating to wider issues concerning morality and social ideals, as well as to rivalry between consumers. Discourses around car use are employed both to sustain wider social ideals and in the creation and maintenance of social difference, through strategies which may again involve delegating responsibility for car use onto others.

In the first meeting of the group with parents, Bill identified people driving into Cambridge as the source of social and environmental problems associated with car use: 'I don't see that we are the problem, it's people coming in that are the problem.' In the second meeting, the tensions between urban and rural ideals lying beneath this view became more apparent: 'It's the country people, if they want nice big homes and big gardens, then they have to pay to use what we pay for.' In the final meeting of the group he explained further:

Bill: I've got this thing about town and country at the moment, it's just my . . . bee in my own bonnet but, umm, we . . . live in towns and we are, between us, providing an enormous number of resources that people in the country need, they come in to us and . . . it's our contributions that gives them cut-price swimming pools, libraries, city centres, security, safety, policing, umm, and I would charge people to come in to Cambridge, and I would have tolls on the way in, for parking and for parking provision and for car . . . everything else, that helps provide all of those things that they come in for . . .

Claire: But aren't they paying for it as well . . .?

Emma: Yes . . .

Claire: . . . I mean, I don't know, but aren't they paying for the swimming pools and things as well?

Bill: Perhaps they are . . .

[Laughter]

Bill: . . . are they paying as much as we are in terms of . . .? We have to live with their cars coming in, and we have to provide car parking for their cars, and we have to live with their pollution as they cut through our streets, and I think that there's a price to be paid for that (A5, 98–124).

Such notions of 'otherness' may be factors in accounting for UK central-government policy. Current policy initiatives such as urban road-user charging (see, for example, DETR 1998) are entirely consistent with the dualism between urban and rural presented here, when in fact such a 'distinction' may no longer be valid (Gillespie and Richardson 2000). The quotation illustrates the ease with which relatively straightforward economic 'solutions' to problems associated with car use are proposed, even though Bill's construction of these problems is complex and multifaceted. The extract also illustrates ways in which other part-icipants' unease over these arguments can similarly be constrained within an economic framework.

Vicky, a woman in her thirties with one young child, offered the group the following views:

> You can certainly tell the difference in the amount of cars on the road during school . . . when the private schools are on holiday, even when the state schools are back, you can still tell there's far less and that must be because people are driving children across the city, just to get to a private school (A1, 1667–1671).

Vicky also wondered why we don't all have smaller cars:

> Vicky: I mean, smaller cars are more environmentally friendly. Why don't we all have small cars? Maybe because you need the bigger space, but some people . . .
> Sanjeev: Yes, I know . . .
> Vicky: . . . with perhaps two people in the family still have an enormous car. Well, why? You know, isn't that just status? (A3, 288–295).

In the latter contribution, Vicky seeks to limit the responsibility of people with smaller cars by blaming only the users of larger cars for the environmental problems associated with car use. However, class mechanisms concerning the creation and maintenance of 'distinction' from people who drive 'enormous' cars are also implicit.

These quotations draw attention to the manner in which discourse about car use is used to suggest opinions which would be much more difficult to articulate directly. Social comment regarding people who live in the commuter belt of villages around Cambridge and inimical remarks about private schools and people who drive large cars have here been rendered more acceptable by being placed within the framework of apparently impersonal discussion about car use.

Negotiations of Ethics

While many participants have taken considerable action to limit and reduce their personal car use there remains a considerable gap between what people do and what they say they would like to do. This is consistent with many other recent studies of attitudes towards car use. Stradling *et al.* (forthcoming) found that one-third of car drivers would like to reduce their car use over the next twelve months, although under a quarter of these felt they were likely to do so. It is often argued that this gap remains because public transport is currently insufficiently appealing to present people with an acceptable alternative to car use (see, for example, Transport 2000 (1997)). Much transport policy seems to be based on a similar premise. The main aim of the recent government White Paper on transport is 'to increase personal choice by improving the alternatives [to the private car] and to secure mobility that is sustainable in the long term' (DETR 1998). Many environmentalists believe that it is a lack of information that constrains agency, responsibility and pro-environmental behaviour (Eden 1993). It would also be easy to conclude from this gap simply that people are hypocritical.

I argue instead that there are plural ethics associated with car use in everyday life, and intense negotiations between these ethical stances. Issues associated with ethical concerns include thriftiness, shopping, work, and caring and concern for others. What will also emerge is the way in which positive benefits that are acknowledged as deriving from car use come up against the increasing awareness of the social and environmental consequences of car ownership and use. These are analogous to contradictions which may be encountered in many others areas of consumption (Miller 1995, forthcoming).

Sanjeev compared the cost of driving into town with the cost of going by bus:

> Sanjeev: [M]y consideration is, economy, the car works out always cheaper than the bus in Cambridge . . .
> Susan: It's sort of perceived to be cheaper, isn't it . . .
> Sanjeev: . . . from my house, four of us have to go, I'll end up paying round about three pound fifty or something. If I take the car I can pay one pound twenty, park there, do the shopping, come home (A2, 477–485).

Susan, a woman in her thirties with three young children, did not have regular access to a car during the day, since her husband used their car to drive to work. She discussed the cost of her husband taking the train to work so that she could use the car, directly linking the costs of both her husband's and her own car use and household budgeting, and making explicit the complex financial rationalities through which the costs of car use and household thrift may be valued:

> Susan: It's like, my husband catches the train . . . if I need the car, then he'll take the train to work and he travels, it's about twelve miles and it costs him three pounds sixty for a return ticket, and that . . . I mean, if you look at the pro rata, and you're looking at how much it costs to go to London [a much longer journey], it's actually very, very expensive . . .
> Bill: Yes . . .
> Susan: . . . and if he does that twice a week, umm, you know, for a month, that's . . . you know, it's over, it's over thirty pounds, so for me to actually have the car twice a week, it costs us thirty pounds and so, I mean, it's cheaper for him to drive to work and, you know, we just . . . if I take him to work I might come back and use the car, you know, it's perceived to be . . . again, it's perception, isn't it, of cost . . . I mean, the running costs of the vehicle . . . so I tend to try and do everything in one day! (A2, 1043–1059).

Susan also discussed the benefits of their car for shopping once they had children:

> I don't think we've actually felt we really needed to have a car until, you know, you've got children whereby the shopping . . . the volume of shopping's that much greater, 'cos we used to just cycle to Sainsbury's and fill a pannier up, you know, fill your bike up with food and . . . or walk back, you know, stuff like that (A2, 1174–1178).

Many participants felt the car was essential for getting to work. Tomas, a man in his fifties, offered a typical statement:

> I work out of town, I work at [. . .], so I have to commute and I don't think in my present job I could use public transport because of pressures of work. It's just, I just can't work eight 'til five. Sometimes I have to start early, sometimes I have to finish late, and if I relied on public transport . . . I just couldn't rely on public transport, so for journeys to and from work, the car is essential unfortunately (B1, 1144–1150).

Other participants discussed the significance of the car for journeys made within work, not just to and from work, illustrating the cultural and social relations which sustain car use at work as well as in wider everyday life. Anne noted the following:

> It's chicken and egg, isn't it, with a car and your job, because . . . because most people are expected to have a car, you're expected to tote all that stuff around with you, and you're expected to arrive, you know, looking neat and tidy and ready to work rather than sort of, you know, peel off your cycle clips and, you know, splosh yourself with water (B5, 1931–1936).

Car use is often associated with a better and a safer social life, enabling people to meet and spend time with a wider range of people outside the home than would be the case without access to a car. As noted earlier, Sanjeev had moved from London with his job four years previously. He said that they go to London once a month, to see friends and do shopping, and 'the car makes it so easy'. In the in-depth group with people aged fifty or over, Catherine, Liz and Anne discussed their reasons for using the car in the evenings:

> Liz: I don't like the idea of coming back in the dark. I mean, I had two or three bad experiences on Parker's Piece [a large open space in the city] at night, nothing absolutely awful but enough to sort of frighten me, to think twice about going across that . . .
> [Others agree]
> Liz: You know . . .
> Catherine: Well, that's why I tend to drive if I'm on my own in the evening, I'd rather . . . sometimes I cycle . . . [but] if I know I'm going to be coming back at ten o'clock, I'd rather, I'd rather be in the car . . .
> Anne: Yes, I'll do the same. If it's Saturday morning, I want to go into town and I've not got too much to carry, then I'm quite happy to walk in, but certainly in the evenings, you park your car in the side streets somewhere and just come back again . . . it does feel safer (B1, 362–379).

Joan, a woman in her seventies who lived alone, shared some of her feelings about her car in the context of her visits to see her niece, who lived in a village about ten miles away:

> Joan: It's a better social life too, isn't it . . .
> Peter: Yes . . .
> Joan: . . . really, you can do more things. I mean, I could never get to [. . .] to my niece for a birthday party, easily, even though I have to go

on the dreaded A14. Had the only accident in forty years on there, just shunted but, umm, I still prefer to, you know, run the risk and try not to go in the rush hour, but, umm, otherwise, I wouldn't . . . well, there is a bus out to there but probably once a day, I mean, I wouldn't be able to do it when I wanted to, would I? (B2, 1525–1536).

These meanings of car use are relatively well-explored, since they are associated with individual personal mobility, safety and control, issues that are strong themes in contemporary discourses around car use. However, even though the above examples concern social life, an emphasis on largely individualistic meanings does not adequately represent the social relations in which car use is embedded. Car use can often be an expression of help, care or love.

This is clearly illustrated by car journeys that are undertaken primarily for the benefit of immediate family members. In the group with parents, participants discussed the benefits of the car for family outings. Emma, a woman in her thirties with two young children, explained why they had driven a round trip of over fifty miles to go swimming:

Well, we wanted to go swimming on Sunday, and the girls wanted to go somewhere they could have fun and go swimming so we drove to Bury [St Edmunds], 'cos there isn't a nice swimming pool in Cambridge (A2, 1356–1359).

Many members of the group with people aged fifty or over were also parents, of course, and some of the younger members still had children at home. Catherine had two teenage daughters and shared her ambivalent feelings towards some of the trips that she had made when they were younger:

Catherine: [O]ne of my regular things was delivering teenage daughters to aerobics classes, which meant me getting the car out, taking them there, coming home and then going out again to collect them again, and I used to think, 'This is stupid' . . .
[Laughter]
Joan: But you can't let them jog home, can you?
Catherine: No . . .
Others: No (B3, 1435–1446).

Although Catherine felt her use of the car was 'stupid', this extract illustrates the social and familial relationships through which meanings of car use are created and maintained. The care for immediate others implicit in these trips, termed 'escort' trips in the dry language of most transport research and policy, was apparent, despite the ambivalence which Catherine expressed.

The care and love implicit in use of the car was most evident in trips made primarily for immediate family members. However, car use also sustains and is sustained by wider social networks. Mary, a woman in her seventies, told the group with people aged fifty or over how she and Joan used their cars:

> I use my car mostly as a taxi service, you know, I constantly take . . . well, [to Joan] you are too, you're constantly taking people backwards and forwards to various places, the elderly and infirm (B2, 488–491).

Later in the same meeting, Mary illustrated the reciprocal nature of these relationships:

> Mary: I don't think it would be the end of the world if I didn't drive any more. I'd use somebody else like they use me . . .
> [Laughter]
> Mary: . . . but I do enjoy it . . .
> Facilitator: Right . . . what do you enjoy about it?
> Mary: Umm, I just enjoy driving. I usually like to have someone by my side, I must admit, if I'm going somewhere that I don't know (B2, 1630–1640).

For Mary, meanings of car use are embedded in the informal social networks through which elderly people help and support each other. In addition, Mary also discusses her enjoyment of driving and, in passing, illustrates how car use is not always only a means to an end, but can also provide directly shared social experiences. Transport policy-makers and environmental campaigners sometimes argue that an emphasis on 'time taken' rather than on 'time spent' and the importance of 'the path as a social space' is relevant only to pedestrian, bicycle and public transport journeys (Krämer-Badoni, 1994), ignoring the fact that journeys by car can also be enjoyable in themselves.

Several members of the group with people aged fifty or over had elderly parents, often living a considerable distance from them. As well as making the journeys to visit them in the first place much easier, trips together in the car were often an important and intimate part of the visit. Anne told the group about her visits to her mother, clearly illustrating once again the profound caring relationships through which meanings of car use are constructed:

> Anne: [I]t would a sad disappointment to her were I to choose to go to visit her by train, because one of the things she really likes is to go, to be taken out for a drive . . .
> Others: Mmm . . .
> Anne: . . . you know, on a Saturday, we'll go out to a pub somewhere and then we'll tootle round the countryside and go back again and, you know, that's one of the features of the visit, so she would miss it a lot (B2, 1668–1677).

A constant need to make choices is a primary feature of modernity. A heightened sense of responsibility may accompany increased choice, and there is therefore a risk of making wrong choices, or choices that may be accompanied by anxiety, perhaps experienced as guilt or shame. However, Warde (1994) argues that the people most likely to experience anxiety associated with consumption develop highly disciplined consumption patterns to displace this anxiety, and the rest are not troubled.

The significant efforts which many participants had made to reduce their personal car use can be viewed as examples of highly disciplined behaviour designed in part to reduce guilt and anxiety, without detracting from the environmental significance of their actions. However, the argument that people who do not develop highly disciplined consumption patterns do not experience guilt or anxiety would seem to be misleading with respect to car use in the light of material presented in this chapter. Despite the changes that many people have been able to make, they may still feel they have not been able to do enough. Others feel that they are unable to make any changes at present. Many meanings of car use in everyday life seem to result from the gap between what people do and what they say they would like to do, and serve fundamentally and perhaps primarily to reduce intense feelings of guilt and anxiety.

The currently dominant approaches to consumption assume far too individualistic a model of the consumer. The material presented here suggests that meanings of car use are fundamentally embedded in social relations of everyday life, and that an understanding of the inter-relationships between the plural ethical discourses associated with car use provides an alternative means of understanding the gap between attitudes and behaviour. In particular, positive social frames of meaning of car use associated with care and love for immediate others, as well as care for others within wider social networks, though fundamental, have been almost completely neglected in academic and policy discussions of car use.

Deliberation and Inclusion of Multiple Meanings of Car Use

Discourses associated with car use are dominated by a limited number of issues and concerns. Arguments which seek to legitimate car use through constructions of lack of choice and agency, delegation of responsibility and promotion of modest personal and individual changes in levels of car use, but not ownership, are widespread both in everyday life and in UK government policy. Economic and technically-orientated 'solutions' to problems associated with car use predominate in many private, individual constructions of meaning associated with car use and in UK government rhetoric and policy.

In contrast, many other meanings of car use are not well-rehearsed, and are 'not immediately available to discourse' (Giddens 1987, cited in Eden 1993). This chapter has explored positive meanings of car use embedded in the social relations of home and family which are only articulated in an extremely fragmented form. Positive emotional meanings of the car are also poorly articulated, and are often rendered difficult to acknowledge by both guilt and embarrassment.

Peter, a participant in his fifties, spoke about the 'emotional attachment' that we have to our cars, admitting that 'in fact, my wife's caught me talking to the car', illustrating in a very open manner an element of his relationship with his car that had not been easy to acknowledge even at home. Neil, another participant in his fifties, discussed his enjoyment of driving. He was careful to locate this enjoyment as an exotic experience, thereby rendering it untypical. The following extract illustrates the difficulties of transcending the well-rehearsed discourses of advertising, and also the humour to which anyone attempting to articulate his or her emotional experiences of car use may be subjected:

> Neil: I do actually enjoy driving, umm, you know, there are times when I choose a holiday where I'm driving and moving along almost every day, I get an enormous thrill, for example, in Spain, driving across the meseta, hour after hour of the same kind of scenery, it induces a kind of hallucination in me . . .
> [Laughter]

Neil: . . . an impression of space and grandeur . . .

Catherine: He's the car ad man!

[Laughter]

Anne: Takes his breath away!

[Laughter]

Neil: . . . I suppose I should feel guilty about that too but, you know, it is only for a week or two each year but, as I say, in the city it's a . . . it's a waste of time (B2, 1584–1607).

Catherine also struggled to articulate some of the emotional meanings of her car to her. Like the extract above, the following discussion occurred in the second meeting of the group with people aged fifty or over. The trusting relationships which had already been established in the group, as well as the space provided by the entire set of meetings, meant that it was possible to explore these intangible issues:

Catherine: I've found there's a difference having my own car. I mean, we've always had a family car which is my husband's work car, and then I got to the point where I had my own, and I did actually feel different, although it was old and second hand, the sense . . . I mean, it wasn't actually just because I could . . . it wasn't just the convenience, because the other one I could use quite a lot of the time, but if I'm honest, there was this little emotional thing . . . that's mine . . .

[Laughter]

Catherine: . . . and I can go exactly where . . . you know, so again, it is that independence thing, and a . . . and also perhaps slightly, just a sense of, you know, it makes you feel like a real person . . .

Joan: Possibly, we give them names, I do . . .

Catherine: . . . well, mine hasn't got a name, but it's definitely mine . . .

Tony: Do you think . . .

Catherine: . . . and also when my husband then gave up his job and lost his car, I suddenly thought, this is going to be the family car now, and I actually felt a bit . . . because I'd bought it, you know, and I had this little feeling of actually, that it was a little bit of my territory, and I think because it's outside . . . you know, [to Facilitator] if you're actually asking what cars mean to people . . .

Facilitator: Mmm

Catherine: . . . because it was outside a house that I share with three other quite large sort of active people, and that this was my space and, you know, even though I don't use it and I don't drive terribly far in it, so there's sort of emotional little . . . (B2, 1744–1778).

There are few opportunities for a wide range of meanings of car use and associated issues and problems to be explored and contested by local people within decision-making processes. As a result, decision-making processes are often felt to be exclusive. Anne felt that one of the principal problems associated with car use was 'the imposition of solutions' onto the local area by decision-making institutions. Referring to people within the local area, Bill felt that 'we've got to decide amongst ourselves where we live whether we want to live with the problems of the car, health problems and all the other associated problems'. Bill acknowledged a 'dichotomy' in his own attitudes towards the car, but argued:

> [T]he ideas have got to come from us, and then they've got to go up to the council, rather than the council deciding what they want in the grand scheme of things, 'cos if they do that, we probably won't end up with a decent provision for [this area] (A3, 947–951).

New deliberative and inclusionary processes of decision-making are required within the contexts of transport and local environmental planning to enable the tensions, ambivalences and contradictions which characterize car use in everyday life to be addressed and worked through, collectively, at a local level.

Conclusions

Positive meanings of car use associated with thriftiness, shopping, work and social relations with immediate others, friends and neighbours coexist with more problematic aspects resulting from the social and environmental consequences of increasing levels of collective car ownership and use. In addition, frames of meaning exist for each of us as pedestrian, cyclist, bus user, parent, shopper or commuter. Miller (forthcoming) argues that contradiction can be avoided if such frames can be kept apart. While people are aware of the different ethics which exist respectively within each such context for car use in everyday social life, and experience guilt and anxiety as they become aware of the contradictions and ambiguities that exist between them, numerous social constructions, as well as the different means of transport themselves, can serve to keep them apart. Exploration of car use through the medium of in-depth groups brings such contradictions into juxtaposition. It also enables some of the connections, contradictions and entanglements of meanings within and between these frames to be articulated more fully, although it is difficult to know the degree to which they are juxtaposed in everyday life outside the particular construction of an in-depth group. However, the evidence highlights not so much a failure of car users to bring together the varied implications and consequences of their relationships to the car, but the often one-sided representation that may be given by decision-makers.

A more honest and open articulation of meanings of car use in everyday life is essential on the part of these decision-makers in their consideration of car users as consumer-citizens if transport policy is to engage more successfully with meanings of car use in everyday life.

References

Bourdieu, P. (1984), *Distinction: a social critique of the judgement of taste*. London: Routledge & Kegan Paul.

Burgess, J., Limb, M. and Harrison, C. (1988a), 'Exploring environmental values through the medium of small groups: 1. Theory and practice'. *Environment and Planning A* 20: 309–26.

—— (1988b), 'Exploring environmental values through the medium of small groups: 2. Illustrations of a group at work'. *Environment and Planning A* 20: 457–76.

Department of the Environment, Transport and the Regions (DETR) (1998), A *new deal for transport: better for everyone*. London: HMSO.

Douglas, M. and Isherwood, B. (1996), *The world of goods: towards an anthropology of consumption*. London: Routledge. First published 1979.

Eden, S. (1993), 'Individual environmental responsibility and its role in public environmentalism'. *Environment and Planning A* 25: 1743–58.

Fine, B. and Leopold, E. (1993), *The world of consumption*. London: Routledge.

Gillespie, A. and Richardson, R. (2000), 'New ways of working: implications for planning and transport'. Paper presented at Royal Geographical Society–Institute of British Geographers Annual Conference, Brighton, 4th–8th January.

Goodwin, P. (ed.) (1995), *Car dependence*. London: RAC Foundation for Motoring and the Environment.

Harrison, C. and Burgess, J. (1994), 'Social constructions of nature: a case study of conflicts over the development of Rainham Marshes'. *Transactions of the Institute of British Geographers* 19, 3: 291–310.

Jackson, P., Brooks, K. and Stevenson, N. (1999), 'Making sense of men's lifestyle magazines'. *Environment and Planning D: Society and Space* 17: 353–68.

Krämer-Badoni, T. (1994), 'Life without the car: an experiment and a plan'. *International Journal of Urban and Regional Research* 18, 2: 347–56.

Miller, D. (1987), *Material culture and mass consumption*. Oxford: Blackwell.

—— (ed.) (1995), *Acknowledging consumption: a review of new studies*. London: Routledge.

—— (forthcoming) (personal communication), 'The dialectics of ethics and identity'. In *The dialectics of shopping (The 1998 Morgan lectures)*. Chicago: University of Chicago Press.

Stradling, S., Meadows, M. and Beatty, S. (forthcoming) (personal communication), 'Helping drivers out of their cars: integrating transport policy and social psychology for sustainable change'. Submitted to *Transport Policy*, November 1999.

Transport 2000 (1997), *Blueprint for quality public transport*. London: Transport 2000 Trust.

Warde, A. (1994), 'Consumption, identity-formation and uncertainty'. *Sociology* 28, 4: 877–98.

The Colonizing Vehicle

Gertrude Stotz

Introduction

During 1987 and 1989 I conducted anthropological fieldwork at Nguru which is one of four Aboriginal Outstation Communities situated approximately 75 km west of Tennant Creek in the Northern Territory of Australia. Nguru[1] is a Warlpiri community that was granted land rights under the Aboriginal Land Rights (Northern Territory) Act 1976 only in June 1990. Between 1987 and 1989 Nguru had an adult population of 25 permanent and approximately another 25 non-permanent residents. There are three or four languages spoken at Nguru: Warlpiri, Warlmanpa, Mudbura, Warumungu, and Aboriginal English as well as plain English.

In late 1985 a bore was sunk and good water found at Nguru. After another year of camping in makeshift shelters in the area, the first six corrugated iron shelters were built and two toilets, a shower and a water tower were erected. At that time the Federal Government (Aboriginal Benefits Trust) provided a Toyota Landcruiser (trayback) and a tractor for communal use. Almost overnight, the people residing at Nguru became a 'Community'. These objects put a tremendous strain on the group. People say that especially since they got the 'Toyota' they have had only trouble. 'Toyota' has become a generic term among Aborigines for all 4WD vehicles.

The Warlpiri exchange system is based on a gendered relationship of rights and obligations to Country and people (see Appendix p.242 for a more detailed explanation). The term *kirda* is used to refer to a relationship through one's patriline; the term *kurdungurlu* is used to refer to a relationship through one's matriline. Although only some individuals have a primary *kirda* or *kurdungurlu* relationship to Nguru

223

Figure 10.1 Hunting Trip, Gnappamila<u>n</u>u, 1989.

itself, the majority stand in primary or secondary *kirda-kurdungurlu* relationship to each other and to Country as their primary *kirda-kurdungurlu* relationship is to other places in the region. Not all these places have been privileged for settlement. Social relations are, however, based on a number of sites, depending mainly on where residents have their Mother's or Father's Country. Site relations also depend on where people were born, where relatives died, where they were initiated, where they used to live, where they may or may not hunt, where and with whom ongoing relationships exist or are being negotiated, with whom they have grown up and, finally, on whether they have access to mobilized transport to maintain and create relationships.

The permanent residents of Nguru are three actual brothers and a sister (the 'bosses', *kirda* – for which term see the Appendix) who are married to three actual sisters and their classificatory brother. Four more married couples constitute the senior and oldest members of the community. There was also a young couple, four widows, a single woman and three bachelors. I lived with the widows and the single women in the *Jilimi*, often glossed as 'single-women's camp'. The non-permanent residents were all considered close of kin to the permanent residents. All together these people made up four intermarrying groups. I was integrated into the local kin group as Nampijinpa, one of eight possible classificatory skin names, and as such I had to learn proper behaviour when it came to gift exchange and proper social conduct. I am still learning.

The Use of the Car

In people's thinking, there are no physical barriers to where a car can go and there are no limits to what a car can endure. People use the car primarily for hunting, shopping, ceremonial travel, visiting family in hospital and jail. The car is a sign of prestige and privilege. The car is also a mobile home and private bedroom; blankets and mattress are stored on seats, doubling as seat covers. Often a gun is placed under the front seats and game is shot from the side window.

During disputes the car is often the object of anger and its windscreen and headlights are smashed with stones or fighting sticks. There seems little regard for the monetary value of such damage. In due course repairs will be done, but until such time the car is used regardless. Engines are 'cooked' (overheated) regularly by older generation drivers, because engine oil, coolant and fuel are being used up to the last drop in similar fashion. The younger-generation drivers maintain cars somewhat more effectively. Nevertheless, the roads on Aboriginal Land are littered with car bodies. Abandoned cars are archaeological sites, for hundreds of metres around them there are discarded parts and rubbish, extinguished camp fires indicating the long wait for a lift or a visit to remove parts. Most often cars are bought from unscrupulous dealers and do not survive the journey back to the Community. Nakamarra said once: 'Since we got Toyota, we only have trouble.'

The aesthetics involved in having country music playing, a cold drink of some sort in one hand and a bag of potato chips or chicken leg in the lap while speeding along some dusty road has to be experienced to be appreciated. But the use and the socialization of the car are not identical matters.

The sitting arrangements are not at all dissimilar to those of the camp. Men never sit face-to-face with women and women sit behind men, although spouses sit beside each other if one of them actually drives the car. The mother-in-law never sits in line with the rear-vision mirror, as her son-in-law might catch her eyes and turn blind as a consequence of transgressing the avoidance barrier. This is to say that, while practical considerations force the women into the back seats because mostly only men can actually drive, kin relations have to be accommodated as well. In fact the socialization of the Toyota requires the accommodation of kinship relations often to the point where this becomes stressful.

As in the camp, children and dogs can sit wherever they choose. Empty bottles and wrappings litter the car, and it is rarely cleaned. This is because the car has an identity (personal communication John von Sturmer). This means that the Toyota has a consciousness of its own, it will 'talk' when it requires the services of 'others'. My classificatory sister Nampijinpa, told me after my car once stalled because I had overheated the engine that 'Your mum got headache, you got to drive slowly'. I was assured that she meant the car.

The Car as a Male Thing

Some features of the car are (seen as) directly related to the traditional power of men over fire (personal communication Marcia Langton). If I were to say any more as to why this is the case, I would transgress into men's business affairs, suffice to say that the mechanical action of the motor is produced by the combustion process; in Warlpiri perception there is a tendency to put the mechanics of the sexual act and other cultural practices such as those pertaining to the car into a one-to-one correspondence. Napananka may have confirmed in part that men have traditionally power over fire when I asked her whether before contact women cooked breakfast for the whole family as they do now, and she replied: 'What for? Men had the fire, no; we just drink ngapa [water] from parraja [wooden container] and go hunting every day, that's all'.

Thus, regardless of why the car might not be working properly, the battery, which is a taboo word in Warlpiri, and the carburetor as well as the starter motor are focused upon when attempts at repairing a stalled vehicle are made. The battery is by far the most privileged source of attention when a malfunction occurs, and it is also one of the first parts of the car to be stolen. Recharging of batteries is assumed to be possible forever, yet batteries are often discarded too soon because recharging facilities rarely exist in the bush. Abandoned cars in the bush are always stripped of their engines and wheels. Before this takes place, however, the carburetor is often taken out of its mounting, as it is the true source of 'fire'. Of course this is not incompatible with the functional logic of the motor.

When I probed further into the question why the car seems to be a men's thing the reply from men was that: 'Women got no licence', 'Women can't fix Toyota', 'Women can drive private car', 'Some women got licence', 'Boss got the car first'. Women's responses almost fully corroborated what the men had to say with the exception that they said: 'Women got to get Toyota too', 'Government got to look after women too, we can learn'.

Further, Nungarrayi has told me that 'Toyota is really like kuyu [meat or animal species], totem, you know.' When she told me this she had a big smile on her face, she was obviously making an analytical point, one that I picked up with interest. Thus Toyotas and other cars are said to die, in a sense, return to the land of the ancestors (compare Young in Chapter Two of this volume). It will continue not only as a reminder of past deeds but it will become a site that is revisited over and over again when the need for car parts or even brake fluids, engine oil and transmission fluid arises. These visits are made by men and these sites are also best known to men (See Glowszewski 1989: 175–6).

Boss and Community

When the outstation was finally established in 1985 after more than five years of shifting from camp to camp, it was the first time ever that the present residents of Nguru were provided with shelter, a tractor, a water pump and a Toyota Landcruiser for their own independent use. However, the provision of this infrastructure was conditional on two things: a) that people had land rights or had lodged a land claim, and b) that people showed reason why they wanted to live permanently on their own traditional country.

It was only the Toyota that could actually replace the loss of mobility people had suffered since they were institutionalized since the late 1940s. Who is the Boss for a Toyota of an Outstation Community is thus of utmost significance. First, it is always a man. Secondly, it is also always a man put forward by the residents themselves. This does not contradict Warlpiri traditions as such; rather it is done in anticip- ation that negotiations will have to be conducted with white men and hence also anticipating only men will be given the Toyota to 'look after community'. Thirdly, the Toyota is given to the Boss in accordance with the assumption that his status as one of the 'traditional owners' coincides with political leadership. Moreover, 'traditional owners' are seen as a group of people identical with the residence group which in turn is supposed to be identical in make-up with the traditional local patrilineal descent group (see Morphy and Morphy 1984; Myers 1986).

The Boss is, however, put into a quasi-broker position by the various Government bodies that deal with Outstation Communities as entities.

Government agents, which includes officers of Aboriginal Institutions, relate to the Boss on the grounds that he is a 'traditional owner' and so assume he is the political leader as well. The Boss carries a dual responsibility. On one hand he has to prove his status in the cross-cultural sphere and on the other hand, on account of the Toyota, he has added responsibilities to provide for and respond to the needs of the Community.

People at Nguru take the idea of Community quite seriously, and not only when it comes to the use of the Toyota. There is a great sense of 'we mob' in relation to the external world, Yapa (Warlpiri person[s]) and Non-Yapa ('whitefella'), of us as bush-mob versus them as town-mob, of us as those of Nguru and those of other Outstations, of this side and the other side – that is, us as kin and them as affines or different mob. But Community in the sense of a politically and economically constituted democratic entity seems a Western imposition (Hamilton 1987; Altman 1987).

Women make an effort to diffuse the pressure on the Toyota to avoid arguments among themselves and with the Boss by using Aboriginal Institutions in town which purport to have responsibilities for the needs

Figure 10.2 Toyota Pictures: Damien, 12 years old.

of women. Women quite naturally want to reclaim some of their control on mobility and space but they cannot do this with the Community Toyota because it is under male control.

The practice of the separate spheres, of women and men going their own way during the day, is strangely affected by the car too. Women and men lose control of knowing the whereabouts of their spouses. Often I had to bring women right up to their husband after a bush trip and was instructed to say to him that she was with me all the time for 'he might be jealous'. Had she been out hunting with other women, and had they walked back into the camp together, as it would have been the case before Toyotas were granted to Outstation Communities, this problem would simply not have arisen. In relation to the Toyota, gender is also about rights and obligations between kin as well as affines. In a word, gender is about exchange relations.

The main issues raised by the presence of the car in the Community are those of conflict situations and new dilemmas.

Although Strathern (1988: 344) ultimately argues that societies based on commodity exchange and those based on gift exchange are 'incommensurable' because gift exchange is 'an inimitable process', Warlpiri people's situation requires them to deal with this paradox, by incorporating the Toyota as a commodity within gift relations.

How the Warlpiri make external things, such as commodities, their own and how the process places gender relations under stress I hope to show in more detail by discussing the socialization of the Toyota – that is, the Community car.

Western Things and Exchange

The problem of external influences poses itself concretely in the form of Western objects, most prominently the Community Toyota. The Toyota is the main object of socialization, and the rule '*kurdungurlu* got to drive Toyota' applies but is hardly ever implemented.

I found that a critical feature of exchange of Western things is the fact that *Western things are gendered* before they enter the social world of the Warlpiri. This explains the need for their socialization, or regendering as this process could also be called.

Further, I need to emphasize here, in the most general sense that all exchanges are governed by an ethics of reciprocity even if the exchange partners are not related as is the case when Yapa exchange with Non-Yapa. This means simply that the one who receives will owe a value or a thing to the (gift)-giver and will return to the latter a similar value or thing at another time. Traditional exchange cycles can span many Warlpiri groups and can take up to 25 years to complete. On the ground of actual sociality at Nguru all people are related to each other and thus *kirda-kurdungurlu* (reciprocity – see Appendix) identifications apply in relation to every exchange partner at Nguru. There is, however, ample evidence throughout the literature that Non-Aborigines are given classificatory skin names in the effort to integrate them into the exchange system. I too was integrated into the local classificatory kinship structure. However, my relationship to the country where Nguru is situated (and this encompasses my relationship to the local patriline) was that of *kirda* by the Warlpiri reckoning and that of *kurdungurlu* by the Warumungu reckoning; in a similar way the Boss also had a dual identity. As I will discuss below, Non-Aboriginal workers who work in Aboriginal Communities – i.e. building houses, etc. – are considered *kurdungurlu*. Who is their *kirda* then, I asked; 'well, they got to have *kirda* somewhere', I was told. The same applies to things. According to the Warlpiri worldview, things too are produced by procreative processes based on social exchange cycles prescribed since time immemorial by the *Jukurpa*, or the 'Law of the Dreaming'.

External and Internal Gendering as a Colonizing Process

By external gendering of objects I mean the way in which Western things have attached to them Western-gendered notions of producing, maintaining, controlling and exchanging things. These are typically premised on the dichotomous opposition between the sexes, which in part constitutes Western gender relations. More significantly, I am concerned with how this gender ideology dictates race relations, that is to say, it dictates by whom and with whom negotiations for Western things are conducted in the cross-cultural sphere. In dealings for the Community Toyota and most infrastructural items, the assumption is that it is natural that they occur not only with men but between men. The Warlpiri male only apparently shares this Western bias for he has very different criteria that guide his actions as a man. There is no reason for him to assume that this external privileging automatically disen-franchises women. But this is exactly what the outcome of cross-cultural sphere negotiations in fact threatens. For even when women are present at such meetings 'their silence is often wrongly taken as agreement', not by the Warlpiri male but by the Government representatives (House of Representatives 1990: 62). Only when the Toyota is actually handed over does this become apparent. What ensues is a conflict-ridden process of socializing the Toyota to transform it into an internally gendered object. This socialization of Western things is only in part achievable and thus remains incomplete as a socially instituted process. This process is engendered by the nature of the car as such and the way in which it enters Warlpiri society. There exists therefore a conflict between two types of social relations, which become most noticeable around the Toyota. This conflict can be seen as the result of an effort to synthesize externally imposed and internally negotiated forms of rights and obligations in relation to the Community Toyota. The conflict arises from an understanding that Government allocations are given to the Community, with the implicit assumption that men are the leaders of (or, more modestly, represent and look after) the Community. Yet, the ultimate focus of conflict is on the figure of the Boss, the man who is largely responsible for cross-cultural negotiations and who also is given the key to the Toyota. That this process is by necessity a gendered process is due to the fact that in Warlpiri society exchange is gendered. External gendering plays such a significant role because the Warlpiri conduct their negotiations from a different gender regime, and the conflict between the two gender regimes – that is the Warlpiri and the Non-Aboriginal – finds expression via Western things and has thus become an internal Warlpiri problem.

There are, of course, Western things other than the Toyota being socialized; money, private cars, clothes, food, tools, information, education etc. But because not all access to Western things is predicated on Land Rights or the establishment of an Outstation, they are not all acquired through cross-cultural negotiations between men. As a consequence their socialization is more indirect. That is to say, the socialization process for things other than the Community Toyota goes through the same internal gendering but without the continuous challenging of one person's (the Boss's) ability to manage the ownership and user rights of the Toyota by the whole Community. Also, this type of gendering has no repercussions in the cross-cultural sphere.

The socializing of Western things at Nguru does not extend to all objects evenly. Those objects which are privileged for socialization are things which most effectively represent a link to hunter-gatherer traditions. The car, the gun and the crowbar are highly significant. For example, Altman's assessment that vehicles are not effectively integrated in the social economy of the hunter and gatherer in Arnhem Land (North Australia) is only partly true for the Warlpiri. He writes:

ownership of a vehicle, even privately (as distinct from communal outstation ownership) does not necessarily result in production leadership. Because motor vehicles are not traditional production implements there are, as yet, no rules that provide guidelines for their control (Altman 1987: 124).

My contention is, however, that the 'rules' become transparent once indigenous gender relations become part of the analysis. The Community Toyota may not fit any form of production processes as Altman's basically economic analysis suggests, but it is nevertheless the most important tool to the people at Nguru. I would go as far as to argue that the car has virtually replaced walking, obviously essential to the traditional production process of hunters and gatherers. No major hunting trip, no firewood collection and no ritual engagement, not to speak of many other activities, were possible without the Toyota or other motorized transport.

Figure 10.3 Toyota Pictures: Damien, 12 years old.

The gun does not evince problems in terms of gender relations as its external gendering largely overlaps with internal practices, while for the crowbar, there is no external gender attached to it and thus socialization is not an issue. Mobilized transport, however, is a totally different matter for it connects itself in every sense to the traditional practices of women and men equally and is fully integral to present-day needs – that is, the need to move from one place to another for social, ritual and economic purposes.

The people at Nguru seem to have come to a silent agreement that the car is a thing naturally under the control of a man from the local patriline. It connects with the idea that mechanical things are male mysteries, something externally produced in the cross-cultural sphere. The unspoken assumptions of male control by both Yapa and Westerners also set the agenda for race relations and support the argument that sexism and racism have something in common.

The Socialization of the Toyota

In order to trace indirectly the 'biography' of the car in a Warlpiri Outstation Community I do not suggest we compile a history of transformations based upon the car as an object changing its identity from gift to commodity or vice versa (Appadurai 1986). Rather it is a story about how the car is humanized with use and thus becomes intimately involved in the colonizing process.

The car makes tracks; it leaves a history of its successes and failures inscribed in the landscape. It is individualized so that it becomes recognizable like persons from a distance. From its movement and direction are identified the purpose and intentions of the people in it, whose identity is advertised as that of the car. For example, people call out when they see or hear the car approaching, 'Nguru Toyota coming'. From the seating arrangement the kin relations of the travellers are made clear, and those factors regulate who can approach it or get a lift. If one catches a glimpse of the driver, who is almost always a man, and whether it is a man or woman who sits beside him, it is possible to predict the subsequent combination of people who will or will not be found riding in the back of the car.

The car is never an exchange article. TV and video also fall into this non-exchangeable category; like the car they are only lent out or borrowed. Lending means that the object will not come back in the same condition as when it was lent. In the case of TV or video, people say, 'they can't look after it, kids will break it'. However, lending cannot be avoided because no one can deny a legitimate request especially when it comes from people one is supposed to look after. The high cost of replacement of TV or video is not perceived as a prominent problem, for such gadgets are notoriously short-lived in anyone's hands. There is, however, a trend to give especially a 'player' (radio with cassette player) as a gift to older people who often do not know how to use these things. Sometimes such gifts are either stored away or handed back straight away with the words 'you keep it'. This arrangement helps the user to ward off any requests to borrow it, because he or she can always refer the request to the owner, saying for example: 'it's not mine, you got to find jaja (grandmother, mother's mother) for that'.

On the other hand, the Toyota has never been freely given in the first place but has been acquired by hard negotiations at many meetings in the cross-cultural sphere. Should a borrower damage the car all hell breaks loose as accusations fly back and forth. Yet when the situation has cooled down men join in fixing the car, and through their joint labour a whole cycle of male rights and obligations in the Toyota starts again. Nakamarra tells me 'Toyota really men's thing 'cause they fix it all the time'. This attitude, while it is appropriated from Western culture, is totally entrenched. It is a direct link to the cross-cultural sphere from where androcentric gender constructs are so easily transmitted and facilitated because it occurs as if naturally between men. This 'Western' attitude is almost splitting women's rights from men's rights in the car and with it the control of social activities that require transport.

Assignment of ownership to one Boss and accumulated ownership rights through use gives the Toyota a composite identity. On one level the Toyota is under the care of one individual and on the other it attracts many relationships through association with people. The contradictory identity of the Toyota stems from this tenuous relationship between individual control and communal use. The Toyota is continuously under pressure to be exchanged but since it is a thing used by the Community, it cannot be exchanged. This is not only because it has through socialization, from its close association with the Boss, acquired traditional characteristics (Samson 1982: 135), but it has become like Country, a site through which relationships of exchange are established/claimed. As such it seems to have become part of the procreative exchange system and used purely for the purpose of creating exchange relationships. The Toyota becomes like Country, of which the *kurdungurlu* (matrilineal descendants) not the *kirda* (patrilineal descendants) are supposed to be the direct beneficiaries (one hunts on mother's country and one paints mother's country, one sings mother's country), hence the dictum '*kurdungurlu* got to drive'. This is to say, *kirda*, those who inherit country through the patriline, represent country; their physical body is the Country. *Kirda* dance, *kirda* are decorated, *kirda* cannot partake of their totemic species hunted on their patrilineal land. In other words, direct benefits can only come from one's mother's country (the descendants of a man's sister): this is why I was told my car was my mother.

This is a complex statement, and it should not be taken to mean that to be the direct beneficiary of the bounty on one's mother's country is in any way applicable to the car despite the similar ruling. For *kurdungurlu* to drive would imply that he or she becomes enslaved to the *kirda*. It is *kirda* who calls the business (traditionally this is ceremonial business) and *kurdungurlu* has to respond to it. This would mean, in practice, that *kurdungurlu* has no discretionary powers over the manner in which the car is to be used, unlike in real business. As mentioned above, *kurdungurlu* controls design, song and performance, *kirda* dances, is painted up and thus represents. The *kurdungurlu* is in the service of the business at hand regardless of whether this occurs in same sex groups or not. In relation to the car, however, this is what the *kirda* has in mind. For a *kurdungurlu* to actually drive the car (which, as mother, is *kirda*) on his or her own would be tantamount to changing his or her identity to that of the Boss, of ignoring the rights of the *kirda*. This would be seen as having turned *worunga* (mad, deaf). It is thus not practical for *kurdungurlu* to drive, but this is the irony inherent in the car and not the ruling. The people at Nguru have recognized this because they never enforce this rule.

The pressure to exchange is based on traditional objectives but is only threatened when the Boss is seen to fail in his duty to 'look after Toyota' or 'Community'. There is apparently no distinction made between Aboriginal and 'Western' requirements of 'looking after'. For example, when the Boss fails to turn up at mortuary rites, which is an increasingly common phenomenon especially among the young and alcoholics (Dussart 1988), his hold on the Toyota is challenged just in the same way as when he fails to pick up the children from school or when he strands the car because he has no money to buy fuel. 'Looking after' is a socially monitored activity; a person who fails in this duty will soon be marginalized and not listened to. That means that while *kirda* rights cannot be taken from anyone, the rights to 'look after Toyota' can be taken over by other *kirda* without altering the relations in the cross-cultural sphere.

Ownership in Warlpiri society can only be claimed for values and things that are in a sense earned. For example, when I kill a goanna on my mother's country I have earned it by hunting it down and killing it. This is because my mother is *kirda* ('boss') for all the bounty on her Country and I, as *kurdungurlu*, have inherited/earned the right to it, for *kirda* is through the Country looking after *kurdungurlu*. If *kirda* requests to hunt on his or her own Country, it is *kurdungurlu* who gives permission and it is *kurdungurlu* again who will get the catch before handing part of it back to *kirda* (see Turner 1989: 147). This practice shows that while *kirda* 'looks after' *kurdungurlu* 'through acts of representing', *kurdungurlu* looks after' *kirda* by 'working' (painting the dancer and singing for the dancers) and by executing the Law.

Like the Toyota, things which have an external source and access to which is 'unearned' remain under constant pressure to be exchanged so that access to them can be earned; for example the pooling of social security moneys at the card game so that it can be exchanged. Such games are hard work and can last for days after which just reward comes to those who stand in close relationship to the winner.

The Toyota, unlike Country, has a temporal identity founded in two facts of its life: that it will become useless one day (discussed below) and that there is a Western fixation on one personality in the figure of the Boss. The Boss has as part of his strategy become agreeable to the idea of leadership invested in him by outsiders to Aboriginal culture. He accepts his responsibility to front up to the 'whitefella' at meetings. Yet, he is also aware that the people expect him only to lead them to a Toyota but not to lead *them*. The Boss would often say to me 'people don't understand government; they only talk in the camp; they never talk at meeting; they are not behind me' thereby indicating that he needs the Community more than they need him. His interest in the Toyota is of course identical with that of his fellow residents but he has the added advantage that he can claim a double legitimacy to this Western object as I have already explained above. The true dilemma the Boss faces is that he can lose the Toyota to another 'traditional owner' ('boss') and with it he would lose his status as Boss. His dilemma becomes the Community's dilemma and one can say the Toyota is like a Trojan horse: it appears simply to be a vehicle of transport, but actually it is an agent of colonization which allows Western notions of leadership to deeply penetrate almost imperceptibly into Warlpiri society, turning the role of the boss into the Western role of leader.

The car's double identity becomes problematic: when it breaks down and needs major repairs or even replacing, or when a person who had used the car (has been travelling in it) and was member of the Community dies. I shall only discuss the first of these. In the first case representations will be made to the relevant institutions (mainly Outstation Resource Centre and Central Land Council) for financial or practical help. The Boss will draw attention to his financial plight by reminding the Community that they 'got to help'. But whenever the car was stranded in this way Community members invariably could not see the sense in 'giving money' when they did not intend to use a car that was useless. Some people would spread rumours about the past behaviour of the Boss. 'He can't look after Toyota properly; he goes town every day; he never there for going to shop; he always late to pick children up from school; he only looks after self; he gives car to anyone in town when he is drunk; we always pay for diesel; he can't talk strong at meeting; all other Outstation get plenty of new Toyota; government got to help too.'

The Boss himself and his wife impressed on me, as their only ally who could spread the word to their side of the story, that 'people never help with diesel; they don't understand about cars; the Government does not help Aborigines in the bush; people in town are much better off and they forget about us.' In this situation the conflict virtually tears the Boss apart. As the Boss will prefer to strand the Toyota at times when he knows there is 'plenty of money', a conflict between privately owned cars and the Community Toyota arises as well. Some nuclear family units will take off in their own cars for 'shopping' and, indeed, for 'filling up petrol'. People will make themselves scarce by taking advantage of the private car, but this strategy is not available to other couples and women in general. Conflict at Nguru has an added local flavour, conflict is again and again focused on the son-in-law and mother-in-law, in the relationship between the Boss (*kirda*) and his wife's mother (*kurdungurlu*). This is because the Boss and his wife are married irregularly. The Boss had a Warumungu mother and a Warlpiri father. He carries two sets of classificatory skin names, which can be construed as him being his wife's uncle, her mother's brother. The mother-in-law, being the daughter of the actual uncle's sister, is thus the main *kurdungurlu* for Nguru. But even if she were not *kurdungurlu*, she could probably claim a secondary *kirda* relationship by patrilateral reckoning. Her mother-in-law status – reciprocal and strict avoidance relationship between her and her son-in-law – would count just as much if she were a female *kirda*. Thus, despite the ruling '*kurdungurlu* got to drive Toyota', it is not because she is *kurdungurlu* but because she is mother-in-law it is impossible for her to openly approach the Boss. She can only voice her displeasure about her son-in-law through her daughter.

The Boss cannot hand the Toyota over to his close brothers and sister who also are 'bosses' (*kirda*) at Nguru because that would mean he is no longer the Boss for the car and this will put him in conflict with his status as reigning *kirda*, being the oldest of four siblings. In a word, the control of the car at the time of entry into the Community is given to a man whom everyone considers Boss. He has traditional rights as *kirda*, and his co-residents hope that he will use the car to help his social career, matching his status of Community Leader with that of his ceremonial status. People want to be proud of each other.

The internal gendering of the Toyota affects the Boss's performance in the cross-cultural sphere to which traditional relationships of ritual performance are being projected. This is where gender systems interface and where the Warlpiri system is being subverted. Hence the socialization of the Toyota as a differential colonizing process in which the

Warlpiri, as subjects of history, participate. This will eventually mean that the Toyota has to change hands. This is a decision the *kurdungurlu* have to make.

The threat to sack the Boss, to take the keys from him, arises at each minor failing. Sometimes it is simply a matter of the car needing new tyres or a new battery, which threatens the Boss's hold on his position. Should the Boss be replaced, his status as ex-Boss is extremely low. In the cross-cultural sphere, he will not be directly consulted and called to meetings by Government agents on matters concerning the Community, be these capital improvements or social services, or exploration permits to mining companies.

This analysis of the 'gift' of a Toyota to an Outstation Community in accordance with an androcentric notion which recognizes only men as 'owners' and Community leaders illustrates the way in which the car is an externally gendered vehicle of invasion.

Government agents perceive internal conflict as disruptive to the Community. At Nguru, white Government personnel have simply made up their own mind as to which of the 'bosses' they want to deal with, and this has created terrible arguments among the three brothers at Nguru. They accuse each other of 'working' behind each other's back, and this is a very serious accusation indeed. It is vital, therefore, to understand how the Boss constructs his identity in the cross-cultural sphere. As a Warlpiri man he will find it natural to negotiate face-to-face with men. The person he negotiates with, usually also a man, operates on the same assumption. Rather than asserting a gender identity based on biological sex, the Warlpiri male will try to establish a gendered exchange relationship with his counterpart. Far from perceiving himself as dependent client, the Warlpiri man sees himself as *kirda* vis-à-vis Government agents who are by this strategy put into the position of *kurdungurlu*. Whitefella who negotiate with Aborigines are seen to 'help us mob' and they 'have *kirda* somewhere'. It is a *kurdungurlu* who has endless supply and apparently unlimited discretion. He thus wants to get the Government agent into a relationship of give and take. This would mean building up from the ground an exchange relationship, which goes beyond the historical reasoning, which says 'government got to help for what they did to us'.

A *kurdungurlu* who has endless supply and who never demands anything in return maintains control of a one-sided exchange situation. At least this is how the situation appears on the surface. Government agents never say why Yapa should get a Toyota for nothing; the debt of the colonizers to Aborigines is never put into words. Specific individuals may be too polite to question the motives for receiving a Toyota. Not only is this part of the strategy to keep the Toyota business among men, it is also impossible to ask people to reciprocate who have for decades made little effort to understand them. There is a lot at stake here. If men were to break the androcentric basis of their privilege they would deny their individual aspirations as well as their inherited right to be seen to be a good Boss. The men deny thereby that they are actually 'working' for 'whitefella' as they are also waiting to be called upon to really 'work'. Already I have overheard women warning their troublesome men that 'whitefella can't get Toyota like we mob, they got to pay for everything'. That is to say that people wonder when Westerners will show their true/false colours. Until such time as this happens the Boss plays along, and competes with other Bosses from different Outstations for this privilege.

A Critical Overview

The impact of Western things on indigenous peoples has been mostly treated incorrectly as the impact of Western technology upon indigenous technology (Ingold 1990). The effect of Western things especially on the material culture of Australian Aborigines has been commented on over decades often in a contemptuous manner with little focus as to the intra-cultural responses, especially in terms of gender relations. However, the fact that Western objects are, as I argue, socialized rather than simply appropriated by the Warlpiri, shows us that they are active, albeit unequal partners in the colonizing process (Sharpe 1952; Taylor 1988).

My short reading of Altman (1987) will here set the scene to investigate the possibility of a gendered view of the colonial process including the introduction of Western things. Altman analyzed social and economic change among the Gunwinggu people of Arnhem Land. His thesis is based on empirical research of economic production from which he reflects on patterns of social change.

He argues that women's and men's economic performances have been influenced differently by the introduction of Western goods. The most significant aspect of this difference was that women's produce, carbohydrates, has almost totally been replaced by market goods. These goods are relatively cheap so that everyone can easily procure them from shops. By contrast, men's meat production has been increased with the introduction of large feral game, such as the buffalo and cattle, and with the introduction of a more effective 'production technology', such as the gun, the car and outboard motors. Overall consumption levels of protein foods and carbohydrates have not changed significantly, if at all, since colonization.

Introduced species and goods have no 'bisniss' (business) attached to them and they cannot be reproduced from within Aboriginal cultural forms of production. This fact has had a major social impact. My own fieldwork concurs with this, and I would go so far as to state that the colonizing impact of Western goods and things is grounded in this lack of attachment to business.

According to Altman, women's produce, which was previously distributed during ceremonies by 'elders', was never the object of trade. Rather, it was given in payment for ceremonial services, such as the *kurdungurlu* provides among the Warlpiri. Hence women's produce as well as meat contributed by men was exchanged for 'social invisibles', by which Altman means ritual obligations. It was the virtual disappearance of trading in spears since the introduction of the gun and large feral animals which has had grave social consequences in that it diminished the power of old men ('elders') in several ways: they no longer control the distribution of goods; they no longer control the marriage market. On the one hand, guns replaced spears, and on the other, any man can hunt meat from feral animals because ritual taboos do not exist for introduced species. This statement is too conservative for the Warlpiri people I observed. While no longer trading spears and flints or spinifex resin during ritual engagements, they now trade bullets, tyres and knowledge about cars, guns and money. In addition, women trade crowbars, blankets, axes and in lieu of money even artefacts produced for the tourist market.

Altman further reasons that, since cash is available to all and the majority of bought food consists of carbohydrates, and because they are cheap compared to fresh meat, they have deflated the value of women's produce and production activity. The change from gathering to shopping has shifted the economic prestige to the Government, which provides the cash. Altman writes:

> Given that people usually procure carbohydrates and tobacco – goods that only 20 years ago were luxuries – it is not surprising that the government that provides cash for these goods is viewed as a source of infinite wealth (1987: 181).

The fact that self-sufficiency of food production has been replaced by market dependency is to a large extent irrefutable. According to Altman,

Figure 10.4 Toyota Pictures: Damien, 12 years old.

the indigenous division of labour between women and men to provide carbohydrates and meat has remained, but meat production as still significant for men's social prestige as hunter, with the help of new 'production technology', has resulted in more equality among men. Unfortunately Altman does not elaborate this point. The cost of carbohydrates has to be compared with the cost not of fresh meat, but with that of guns, bullets, cars and boats. But I assume Altman to mean that women's contributions are now more than ever accessible by *all* men rather than by some as in the past. If this is the case the implication of social change are too enormous to contemplate. Such a total loss of social prestige for women cannot be envisaged.

Altman's analysis cannot explain how men's privileged access to new technology, which aided the partial preservation of a hunter's life style for them, has resulted in men's economic prestige (in such a manner that) the free access to carbohydrates (shops) by women has created a loss of prestige. This analysis typically ignores Aborigines' point of view and imposes a Western economic interpretation of socio-economic change on Gunwinggu society. This attitude risks missing the enormous contribution that the people themselves make to the present situation.

The economist's point of view is limited and, however great the effort to do justice to socio-cultural criteria, it fails. This, I think, is why Altman is unable to explain why men should for example monopolize the car and what the communal use of large infrastructure items might mean to the indigenous gender system and exchange relationships generally.

What I have tried to show is how the Community Toyota (4WD) proceeds to revolutionize Warlpiri society by subverting the procreative model of exchange and, more fundamentally, by undermining gender relations.

Appendix: Short Introduction to the *Kirda – Kurdungurlu* System

The Warlpiri gender model is at the basis of the indigenous exchange structure. Exchange among the Warlpiri is based on a procreative model of reproduction.

Inheritance of Country, marriage rules and ritual structure are all premised on the fact that one descends from a mother and a father respectively. From father one inherits the relationship *kirda* to one's father's father's Country and from mother one inherits the relationship *kurdungurlu* to one's mother's father's Country. Thus, while ownership of Country is transmitted through the patriline, economic, social and ritual structures are regulated by descent and kinship from an exogamous brother–sister pair.

One is *kirda* for one Country, father's Country, and *kurdungurlu* for another Country, mother's Country. As *kirda* one *represents* the Law (Jukurpa), as *kurdungurlu* one *executes* the Law. The Warlpiri call the *kirda* 'bosses' and the *kurdungurlu* 'workers' or they call them 'kings' and 'lawyers'. The latter is the more adequate description of the relationship.

The *kirda–kurdungurlu* relationship structure represents the procreative model of social reproduction and hence it stands also for the indigenous gender system. Gender system because the sister of a man is equally *kirda* to her father's Country, and a brother of a woman is equally *kurdungurlu* to his mother's Country. A sister represents as aunt (brother's sister) the patrilineal role vis-à-vis her nieces in ritual, the brother represents as uncle (mother's brother) the matrilineal role vis-à-vis his nephews in ritual. Sister and brother, as the exogamous unit, are respectively female fathers and male mothers to each other's opposite-sex children in relationship to their common patrilineal and matrilineal Country. This is to say, as mother's brother, a man is uncle only to his sister's children, the same way as a brother's sister, a woman, is aunt only to her brother's children. When the brother or the sister relate to their respective nieces and nephews, the man relates to them as male mother when dealing with his mother's patriline and the woman relates to them as female mother when dealing with her father's country (see also Malinowski 1962: 163; Weiner 1979: 331).

The Warlpiri conduct all their affairs almost exclusively in same-sex groups; in separate spheres, there is 'women's business' and 'men's business', yet within each separate-sex sphere people relate and engage in exchange according to the above gender system. However, there is only one Law, one Jukurpa, interpreted from different angles for different purposes by women and men. Overall the same procreative model of symbolic – that is, social – reproduction is the aim of every individual's actions.

In relationship to the Community Toyota, it is this relationship that is being subverted and distorted, thereby creating social stress.

References

Altman, J.C (1987), *Hunter-Gatherers Today: an Aboriginal Economy in North Australia*. Canberra: AIAS.

Appadurai, A. (ed.) (1986), *The Social Life of Things, Commodities in Cultural Perspectives*. Cambridge: Cambridge University Press.

Dussart, F. (1988), *Warlpiri Women's Yawulyu Ceremonies: A Forum For Socialization and Innovation*. (Unpublished PhD Thesis, Australian National University, Canberra.)

Glowszewski, B. (1989), *Les rêveurs du désert: aborigines d'Australie, les Warlpiri*, Paris: Plon.

Hamilton, A. (1987), 'Equals to whom?: visions of destiny and the Aboriginal aristocracy.' *Mankind* 17, 2: 129–39.

House of Representatives (1990), '*Our Future, Our Selves:* Aboriginal and Torres Strait Islander Community Control Management and Resources'. House of Representatives Standing Committee on Aboriginal Affairs. Canberra: Australian Government Publishing Service.

Howard, M. (1977), 'Aboriginal political change in an urban setting: the NACC elections in Perth'. R.M. Berndt (ed.), *Aborigines & change: Australia in the 70s*. [placename], N.J.: Humanities Press.

Ingold, T. (1990), 'Society, Nature and the Concept of Technology', *Archaeological Review from Cambridge,* 9, 1: 5–17.

Malinowski, B. (1967), *Sex, Culture, and Myth,* London: Mayflower Books.

Morphy, F. and Morphy, H. (1984), 'Owners, Managers, and Ideology: A Comparative Analysis', *Oceania Monographs* 27: 46–66.

Myers, F.R. (1986), 'The Politics of Representation: Anthropological Discourse and Australian Aborigines.' *American Ethnologist,* 13, February: 138–53.

Sharpe, L. (1952), 'Steel Axes for Stone-Age Australians', *Human Organization* 2, 2: 17–22.

Stotz, G. (1993), '*Kurdungurlu* Got to Drive Toyota: Differential Colonizing Process among the Warlpiri', (Unpublished PhD Thesis, Deakin University.)

Strathern, M. (1988), *The Gender of the Gift.* Berkeley: University of California Press.

Taylor, J. (1988), 'Goods and Gods, A Follow-Up Study of "Steel Axes for Stone-Age Asutralians"', in Swain and Rose (eds.), *Aboriginal Australians and Christian Missions,* Australian Association for the Study of Religions at the South Australian College of Advanced Education.

Turner, D.H. (1989), *Return to Eden: A Journey through the Promised Landscape of Amagalyuagba.* New York: Peter Lang.

Weiner, A. (1979), 'Trobriand Kinship from another view: the reproductive power of women and men'. *MAN* (n.s.) 14, 2: 328–48.

Note

1. Nguru is not the actual name of the Community; also Warlu, which I mention in this chapter, is not an actual name.

Index